T0155576

SpringerBriefs in Speech Technology

Studies in Speech Signal Processing, Natural Language Understanding, and Machine Learning

Series Editor:
Amy Neustein

SpringerBriefs present concise summaries of cutting-edge research and practical applications across a wide spectrum of fields. Featuring compact volumes of 50 to 125 pages, the series covers a range of content from professional to academic. Typical topics might include:

- A timely report of state-of-the art analytical techniques
- A bridge between new research results, as published in journal articles, and a contextual literature review
- A snapshot of a hot or emerging topic
- An in-depth case study or clinical example
- A presentation of core concepts that students must understand in order to make independent contributions

Briefs are characterized by fast, global electronic dissemination, standard publishing contracts, standardized manuscript preparation and formatting guidelines, and expedited production schedules.

The goal of the **SpringerBriefs in Speech Technology** series is to serve as an important reference guide for speech developers, system designers, speech engineers and other professionals in academia, government and the private sector. To accomplish this task, the series will showcase the latest findings in speech technology, ranging from a comparative analysis of contemporary methods of speech parameterization to recent advances in commercial deployment of spoken dialog systems.

More information about this series at http://www.springer.com/series/10043

K. Sreenivasa Rao • N. P. Narendra

Source Modeling Techniques for Quality Enhancement in Statistical Parametric Speech Synthesis

K. Sreenivasa Rao
Department of Computer Science and Engineering
Indian Institute of Technology Kharagpur
Kharagpur, West Bengal, India

N. P. Narendra
Aalto University
Espoo, Finland

ISSN 2191-737X ISSN 2191-7388 (electronic)
SpringerBriefs in Speech Technology
ISBN 978-3-030-02758-2 ISBN 978-3-030-02759-9 (eBook)
https://doi.org/10.1007/978-3-030-02759-9

Library of Congress Control Number: 2018959748

This Springer imprint is published by the registered company Springer Nature Switzerland AG
The registered company address is: Gewerbestrasse 11, 6330 Cham, Switzerland

Preface

Speech is the most natural way for humans to communicate with each other. Synthesis of artificial human speech provides efficient human-computer communication. Nowadays, the speech synthesis systems are widely used in various applications such as screen readers for visually challenged people, speech interface for mobile devices, navigation, and personal guidance gadgets. As humans are very sensitive in perceiving even the slightest distortions in the speech signal, speech synthesizers with suboptimal quality make them unfit for usage in commercial applications. The main goal of this book is to improve the quality of statistical parametric speech synthesis (SPSS) by efficiently modeling the source or excitation signal. The excitation signal used in synthesis should preserve all natural variations so that the synthesized speech is close to natural quality. The work presented in this book confines its scope to the (1) accurate estimation of pitch (F_0) and (2) precise modeling of excitation signal. For modeling the excitation signal, both parametric and hybrid approaches are explored. In this work, creaky voice has been synthesized at appropriate places by proposing appropriate methods and models.

The contents of the book are useful for both researchers and system developers. For researchers, the book will be useful for knowing the current state-of-the-art excitation source models for SPSS and further refining the source models to incorporate the realistic semantics present in the text. For system developers, the book will be useful to integrate the sophisticated excitation source models mentioned in the book to the latest models of mobile/smart phones. The book has been organized as follows:

- Chapter 1 introduces the topic of text-to-speech synthesis. Different speech synthesis approaches are briefly discussed.
- Chapter 2 provides a review of the state-of-the-art methods for F0 estimation and parametric and hybrid source modeling approaches.
- Chapter 3 discusses the design of a voicing detection and F_0 estimation method by adaptively choosing appropriate window size for zero-frequency filtering method.
- Chapter 4 presents two parametric source modeling methods.

- Chapter 5 describes two proposed hybrid methods of modeling the excitation signal.
- Chapter 6 deals with the generation of creaky voice by addressing two main issues, namely, automatic detection of creaky voice and source modeling of creaky voice.
- Chapter 7 summarizes the contributions of the book along with some important conclusions. Directions toward the scope for possible future work are also discussed.

We would especially like to thank all professors of the Department of Computer Science and Engineering, IIT Kharagpur, for their consistent support during the course of editing and organization of the book. Special thanks to our colleagues at Indian Institute of Technology, Kharagpur, India, for their cooperation to carry out the work. We are grateful to our parents and family members for their constant support and encouragement. Finally, we thank all our friends and well-wishers.

Kharagpur, India K. Sreenivasa Rao
Espoo, Finland N. P. Narendra

Contents

Acronyms

AC	AutoCorrelation
AMDF	Average Magnitude Difference Function
APP	Aperiodicity, Periodicity and Pitch
CART	Classification And Regression Tree
CC	Cross Correlation
CD	Continuous probability Distribution
CMOS	Comparative Mean Opinion Scores
CRD	Cumulative Relative Dispersion
CSTR	Centre for Speech Technology Research
CW	Characteristic Waveform
DNN	Deep Neural Networks
DSM	Deterministic plus Stochastic Model
EGG	ElectroGlottoGraph
EM	Expectation-Maximization
ESPS	Entropic Signal Processing System
F	Female
FPE	Fine Pitch Error
FPR	False Positive Rate
GCI	Glottal Closure Instants
GPE	Gross Pitch Error
HE	Hilbert Envelope
HMM	Hidden Markov Model
HNR	Harmonic-to-Noise Ratio
HTS	HMM-based Speech Synthesis System
Hz	Hertz
IAIF	Iterative Adaptive Inverse Filtering
IFAS	Instantaneous Frequency Amplitude Spectrum
IFP	IntraFrame Periodicity
IPS	InterPulse Similarity
KB	KiloByte
KD	Kane-Drugman

kHz	Kilo Hertz
LF	Liljencrants-Fant
LP	Linear Prediction
LPC	Linear Predictive Coding
LSD	Log-Spectral Distance
LSF	Line-Spectral Frequencies
M	Male
MB	MegaByte
MBROLA	Multi-Band Resynthesis OverLap and Add
MDL	Minimum Description Length
MELP	Mixed Excitation Linear Prediction
MFCC	Mel Frequency Cepstral Coefficients
MGC	Mel-Generalized Cepstrum
MGLSA	Mel-Generalized Log Spectral Approximation
MLSA	Mel Log Spectral Approximation
ms	Milli Seconds
MSD	Multi-Space prabability Distribution
MSE	Maximum Strength of Excitation
NN	Neural Networks
NPI	Next Phone Identity
PCA	Principal Component Analysis
PI	Phone Identity
PP	Position of Phrase
PPI	Previous Phone Identity
PS	Position of Syllable
PSOLA	Pitch Synchronous OverLap and Add
PW	Position of Word
RAPT	Robust Algorithm for Pitch Tracking
RTSE	Relative Time Squared Error
SHS	SubHarmonic Summation
SPSS	Statistical Parametric Speech Synthesis
SRH	Summation of Residual Harmonics
STRAIGHT	Speech Transformation and Representation using Adaptive Interpolation of weiGHTed spectrum
TD-PSOLA	Time-Domain Pitch Synchronous Overlap and Add
TPR	True Positive Rate
UV	Unvoicing
V	Voicing
VDE	Voicing Decision Error
WI	Waveform Interpolation
ZFF	Zero-Frequency Filtering
ZFFHE	ZFF with Hilbert Envelope
ZFFUW	ZFF with Uniform Window
ZFR	Zero-Frequency Resonator

Chapter 1
Introduction

1.1 Introduction

Speech is one of the most natural ways for humans to communicate with each other. Recently, numerous attempts have been carried out to introduce speech interfaces into human-computer communication environments. The speech interfaces are currently used in several portable electronic devices such as mobile phones, household devices, assistive aids for visually challenged people, navigation, and personal guidance gadgets. To develop the high-quality speech interfaces, speech processing technologies in different areas such as speech recognition, dialogue processing, speech understanding, natural language processing, and speech synthesis are very much essential. Text-to-speech synthesis (TTS) is one of the key speech processing technologies used for transmitting information from a machine to a person by voice. The TTS system converts a given text into its corresponding spoken form. For effective information transmission, the TTS system should have an ability to generate natural and intelligible speech. Even though many synthesis approaches have been proposed in the past, the quality of synthetic speech is yet to fully match the quality of human speech.

The high-level block diagram of a speech synthesizer is shown in Fig. 1.1. The TTS system is composed of two modules: (1) text and linguistic analysis (*front-end*) and (2) waveform generation (*back-end*). The front-end is logically separated from the back-end. The front-end identifies abbreviations, punctuations, and acronyms in the input text and converts them into a standard written text. By using the letter to sound rules, every word is converted into corresponding phonetic transcription which describes how the text should be pronounced. Finally, the information regarding the prosodic properties of the utterance such as intonation, duration, and pause patterns are generated. The back-end generates the speech waveform based on the information provided by the front-end. The back-end depends on the type

K. S. Rao, N. P. Narendra, *Source Modeling Techniques for Quality Enhancement in Statistical Parametric Speech Synthesis*, SpringerBriefs in Speech Technology,
https://doi.org/10.1007/978-3-030-02759-9_1

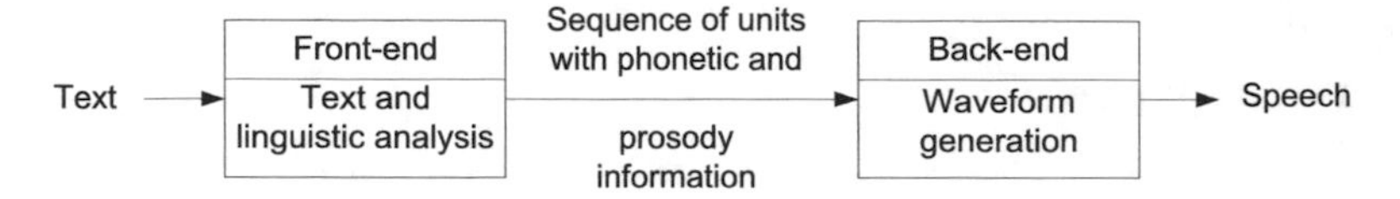

Fig. 1.1 Functional block diagram of a TTS system

of the method used to generate the actual speech waveform, such as concatenation of recorded speech waveforms, and utilize the statistical models and the source-filter-based vocoder. Even though the back-end relies on the information provided by the front-end, the naturalness and intelligibility of the synthesized speech depend on the type and implementation of the back-end. In the following section, different synthesis approaches used in the back-end are described in detail.

1.2 Speech Synthesis Methods

Generally, the parameters and resources used at the front-end are language specific, and the methods and models used in both front-end and back-end are often language independent. Compared to the front-end, the back-end synthesis methods highly influence the segmental speech quality and various voice characteristics of the synthesized speech. Hence, TTS systems are classified based on the type of synthesis methods used in the back-end.

1.2.1 Formant Synthesis

Formant synthesis formulates a set of rules on the acoustic parameters to generate the speech. The acoustic parameters specify formant frequencies, their amplitudes, and bandwidths as well as fundamental frequency, voicing, and the amount of aspiration noise. The human experts derive a set of rules based on the analysis of the speech data. The rules specify how the parameters should vary with respect to time depending on the input from the front-end. The excitation signal which is represented as a sequence of pulses or noise is passed through a synthesis filter which represents the formant frequencies [1]. The synthesis filter is constructed by using several second-order filters where each filter represents one formant frequency. The second-order filters can be connected either in cascade or in parallel. The naturalness of the formant synthesis is poor due to the limited set of rules formulated by the human experts and due to the simple excitation scheme. In practice, it is very difficult to develop a comprehensive set of rules to generate a high-quality speech. In spite of many drawbacks, the formant synthesizer can be used in reading machine for the visually challenged people and are suitable for applications which require small memory footprint [2–4].

1.2.2 Articulatory Synthesis

Articulatory synthesis aims to model the speech in terms of articulatory features of speech production mechanism. They describe the vocal-tract shape and the movement of articulators of speech production mechanism with time [5, 6]. The main issue in articulatory synthesis is the complexity in deriving articulatory rules to produce speech sounds [7]. The existing articulatory synthesizers can produce good quality speech for isolated sounds, such as vowels [8]. But the speech quality is significantly degraded during synthesis of continuous speech, due to the problems in modeling the coarticulation effects. Despite the progress in research on basic speech production mechanism in recent years, the articulatory synthesis has not achieved much success compared to other speech synthesis methods. Recently, the articulatory features derived from the recent measurement techniques, such as magnetic resonance imaging and electromagnetic articulography, have been used in current statistical parametric speech synthesizers [9, 10].

1.2.3 Concatenative Synthesis

In concatenative synthesis, pre-recorded natural speech waveforms are split into small speech segments, and during synthesis, these segments are concatenated smoothly to generate the speech utterances. Generally, the concatenative synthesis provides high-quality speech output, but sometimes audible distortions may occur in the output due to selection and concatenation of inappropriate speech segments. Depending on the type of units used for concatenation, there are mainly three types of concatenative synthesis, namely, (1) domain-specific synthesis, (2) diphone synthesis, and (3) unit selection synthesis. In domain-specific synthesis, the speech utterance is generated by concatenating the recorded words and phrases. This type of approach can be used in applications, such as scheduled announcements in public transportation and weather reports where the desired speech output is limited to a small specific domain. The synthesized speech quality is very high due to the concatenation of long natural speech segments. But the main drawback of this method is that the prosody of the synthesized speech does not match the prosody of the natural speech utterance.

In diphone synthesis [11], a speech database is carefully prepared such that each of the diphones appears at least once and the number of diphones depends on the language. During synthesis, the speech is obtained by joining the diphones selected from the small speech database. The prosody characteristics of the selected diphones are modified using suitable signal processing techniques such as pitch-synchronous overlap-add (PSOLA) [12], multiband resynthesis overlap-add (MBROLA) [11], and residual-excited linear prediction (RELP) [13]. The diphone synthesis frequently suffers from audible distortions when two diphones that are not compatible with each other are concatenated. The audible distortions can also occur during the prosody modification of diphones by signal processing techniques.

Unit selection synthesis generates the speech by concatenating the natural speech segments selected from a large database [14]. The large speech database contains multiple instances of each unit with varying context and prosodic situations. The speech database is segmented into units that can be of variable length: half-phones, phones, diphones, triphones, demisyllables, syllables, morphemes, words, phrases, or even sentences. The choice of the unit depends on the nature of language and the target application. After the segmentation of speech database, the speech units are indexed, clustered, and labeled according to linguistic and acoustic features. During synthesis, depending on the target units and their associated phonetic, contextual, and prosodic features, the suitable units are selected from the database and then concatenated to generate the speech. The unit selection synthesis can provide highly natural and intelligible speech if a large well-optimized (single speaker) corpus is used. The unit selection synthesis is best suitable for applications which are involved in the generation of speech with single speaking style. The drawback of this synthesis is that it is inherently inflexible and difficult to change the voice characteristics such as voice quality, speaking style, and expression.

1.2.4 Statistical Parametric Speech Synthesis

Statistical parametric speech synthesis (SPSS) uses the principle similar to formant synthesis which involves parameterization of speech during the training phase and reconstructing the speech from the parameters during the synthesis phase. In the SPSS, the speech is parameterized based on source-filter representation. The source refers to the excitation signal produced due to the vibration of vocal folds during voiced speech and the release of constriction at different places along the vocal-tract during unvoiced speech. The filter refers to the cascade of resonators used for realizing the shape of the vocal-tract system. In general, the vocal-tract filter is represented by spectral envelope of the speech signal. The spectral envelope of the speech signal and the excitation signal is parameterized and modeled by the hidden Markov models (HMMs) in a unified framework. As SPSS uses HMMs for modeling the parameters, it is often called HMM-based speech synthesis system (HTS). However, recently deep neural networks (DNN) have also been used for modeling the parameters in SPSS [15, 16].

As the SPSS uses parametric approach, the generated speech is smooth and intelligible. The SPSS has higher parametric flexibility compared to unit selection synthesis, and it can be adapted [17] to a different voice quality, speaking style, and emotion by using a small amount of target speech data. The amount of speech data required for training the SPSS is very small compared to the unit selection methods, and the memory footprint of SPSS is also very small. The degradation in the quality of speech synthesized from SPSS is mainly due to two factors [18], namely, (1) buzziness and (2) muffledness. The buzziness is caused due to improper parameterization and modeling of the excitation signal. The muffledness is caused by over-smoothing of the generated speech parameters due to averaging in the statistical modeling. Recently, several improvements have been proposed to improve the quality of SPSS, which makes it acceptable even for commercial applications.

1.2.5 Hybrid Synthesis Methods

Hybrid synthesis methods are proposed to combine the advantages of statistical approach with the naturalness obtained from the unit selection. The HMM-based approach has been used in unit selection synthesis to predict target units and computing cost functions [19–21]. The hybrid synthesis methods can also use the HMM-based approach as a probabilistic smoother of the spectrum across speech unit boundaries [22, 23]. Multiform speech synthesis has been proposed which mixes the natural and statistically generated speech frames to improve the perceptual quality of the synthesized speech [24–27].

The hybrid synthesis methods have several advantages compared to unit selection and HMM-based synthesis methods. The buzziness and muffledness occurring in the synthesized speech of SPSS are reduced by using the natural units. The HMM-based cost functions assist in the selection of suitable units with appropriate context. However, the hybrid approaches lose the benefit of flexibility and small memory footprint by using the natural units.

1.3 Objectives and Scope of the Work

The source or excitation signal used for synthesis significantly influences the quality of speech. The excitation signal should preserve all the natural variations so that the synthesized speech is close to natural quality. One of the important attributes of the excitation signal is the pitch or fundamental frequency (F_0). Generally, the performance of existing voicing detection and F_0 estimation methods is high in modal voiced regions, but their performance degrades sharply in creaky and low-voiced regions. To address this issue, a robust voicing detection and F_0 estimation method are proposed by using zero frequency filtering (ZFF) method [28]. The size of the window used in ZFF is exploited for accurate voicing detection and F_0 estimation.

The modeling of the excitation signal is a very critical issue in HMM-based speech synthesis. Initially, the parametric approach of modeling the excitation signal is explored. In the parametric approach, the excitation signal is decomposed into a number of pitch-synchronous residual frames. The pitch-synchronous residual frames are GCI centered and two pitch period long excitation segments. The pitch-synchronous residual frames are accurately represented in terms deterministic and noise components. Here, parameterization is performed based on the analysis of characteristics of residual frames around glottal closure instants (GCIs). Later it is observed that the parameterization and statistical modeling tend to over-smooth the finer detailed structure of the excitation signal. Hence, to preserve the natural variations, and to accurately represent each of the residual frames, a hybrid source modeling approach is proposed. In the hybrid approach, a combination of natural waveforms and parameters is utilized for modeling the excitation signal. During modeling, the phone-dependent characteristics of the excitation signal have been given appropriate importance.

Generally, most of the source modeling methods are optimized to generate speech corresponding to modal phonation [29–31]. But synthesis of creaky voice is necessary, as the humans involuntarily produce creaky voice at certain places in the speech utterance. To generate creaky voice, first the creaky regions should be identified in the speech utterance. In this work, a creaky voice detection method is proposed by analyzing the variation of epoch parameters in modal and creaky regions. After the identification of the creaky regions, the excitation signal corresponding creaky voice needs to be modeled. Due to the presence of secondary excitation in addition to primary excitation at GCI in the creaky excitation signal, the method used for modeling the modal phonation cannot be used directly for modeling the creaky excitation signal. Hence, an appropriate creaky source modeling method is proposed by extending the hybrid source modeling meant for generating modal voice.

1.4 Contributions of the Book

The major contributions of this work include (1) robust voicing detection and F_0 estimation, (2) exploring both parametric and hybrid approaches for modeling the source signal, and (3) automatic detection of creaky voice from the speech utterance and source modeling for creaky voice. The brief overview of each of these contributions is discussed in the following subsections.

1.4.1 Robust Voicing Detection and F_0 Estimation Method

The performance of current voicing detection and F_0 estimation methods [32–34] used in HMM-based speech synthesis system degrades sharply for creaky or low voicing regions. In this work, a robust voicing detection and F_0 estimation method is proposed using zero frequency filter method. The ZFF method is widely used to derive the locations of impulse excitation. The size of the window used in ZFF is exploited for accurate voicing detection and F_0 estimation. The performance of the proposed method is compared with other existing voicing detection and F_0 estimation methods. The proposed method is used for modeling the F_0 patterns in HMM-based speech synthesis system. Both objective and subjective evaluation results show that the proposed method is capable of generating good quality speech compared to two other existing voicing detection and F_0 estimation methods.

1.4.2 Parametric Approach of Modeling the Excitation Signal

In the parametric approach, two methods are proposed for modeling the excitation signal. In the first method, principal component analysis (PCA) is performed on the pitch-synchronous residual frames extracted from the excitation signal. Based on the analysis, every pitch-synchronous residual frame is parameterized using 30 PCA coefficients. In the second method, an analysis of characteristics of the residual frames around GCIs is performed using principal component analysis. Based on the analysis of the residual frames around GCI, the pitch-synchronous residual frames are decomposed into deterministic and noise components. Both deterministic and noise components are parameterized and modeled using HMMs. During synthesis, the excitation signal is reconstructed from the parameters generated from HMMs. Subjective evaluation results show a significant improvement in the quality of the proposed method, compared to three existing source modeling methods.

1.4.3 Hybrid Approach of Modeling the Excitation Signal

In the hybrid approach, initially, a source modeling method is proposed where the excitation signal of every phone is modeled by using a single optimal residual frame. During synthesis, for the given input phone, the excitation signal of a phone is constructed from the suitable optimal residual frame selected from the database. To further enhance the source modeling method for generating the excitation signal close to the natural quality, the time-domain deterministic plus noise model-based source model is proposed. In the proposed hybrid source model, phone-specific characteristics of the residual frames are included in modeling and generation of the excitation signal. The pitch-synchronous residual frames of a phone are modeled as a sum of deterministic and noise components. The deterministic components estimated from all phones are stored in the database. The noise components are parameterized and modeled using HMMs. Performance evaluation results show that the proposed method is capable of producing natural sounding synthetic speech which is clearly better than the state-of-the-art HMM-based speech synthesis systems.

1.4.4 Generation of Creaky Voice

Generation of creaky voice is carried out by addressing two main issues, namely, automatic prediction of creaky voice and source modeling of creaky voice. An automatic creaky voice detection method is proposed based on the analysis of variation of epoch parameters for different voicing regions. Generation and modeling of the creaky excitation signal are performed by extending the proposed hybrid source modeling approach. In the proposed hybrid source model, the creaky excitation is generated as a combination of creaky deterministic and noise components. Subjective evaluation results indicate that the incorporation of creaky voice at appropriate places in the synthesized speech has improved its naturalness and overall quality.

1.5 Organization of the Book

The book is organized as follows:

- **Chapter 1: Introduction** introduces the topic of text-to-speech synthesis. Different speech synthesis approaches are briefly discussed. The objective and scope of this book and overview of contributions are briefly discussed.
- **Chapter 2: Background and literature review** briefly discusses the HMM-based speech synthesis system. State-of-the-art methods for F0 estimation, parametric, and hybrid source modeling approaches are briefly discussed. The previous works carried out for the detection of the creaky voice and the modeling of the creaky excitation signal are also described. In this chapter, the limitations of the state-of-the-art methods are provided, and the motivation for the proposed method has been derived from the limitations of the existing methods.
- **Chapter 3: Robust voicing detection and F_0 estimation method** discuss the proposed voicing detection and F_0 estimation method by adaptively choosing appropriate window size for ZFF method. The performance of proposed method is compared with the existing F_0 estimation methods. The speech synthesized using the proposed method is compared with the speech synthesized using the two existing F_0 estimation methods.
- **Chapter 4: Parametric approach of modeling the source signal** presents two methods of parameterizing the excitation signal. In the first method, the residual frames are parameterized by using PCA coefficients. The second method parameterized the residual frames as a combination of deterministic and noise components. In this chapter, an analysis of characteristics of residual frames around GCI is performed.
- **Chapter 5: Hybrid approach of modeling the source signal** presents two proposed hybrid methods of modeling the excitation signal. In the first method, the optimal residual frames extracted from the excitation signals of phones are used for modeling the excitation signal. The second hybrid source modeling method is based on time-domain deterministic plus noise model. In this model, the waveforms of the deterministic components are stored in the database, and the noise components are represented by parameters. Analysis of phone-dependent characteristics of the excitation signal is also performed in this chapter.
- **Chapter 6: Generation of creaky voice** deals with the synthesis of creaky voice by addressing two main issues, namely, automatic detection of creaky voice and source modeling of creaky voice. The creaky voice detection method is

proposed based on the analysis of epoch parameters in different voicing regions. Modeling of the creaky excitation signal is proposed by extending the time-domain deterministic plus noise model-based hybrid source model.

- **Chapter 7: Summary and conclusions** summarize the contributions of this book along with some important conclusions. This chapter also provides some directions for further research.

References

1. W. Lawrence, The synthesis of speech from signals which have a low information rate, in *Communication Theory*, ed. by W. Jackson (Butterworth & Co, London, 1953), pp. 460–469
2. J.M. Pickett, *The Acoustics of Speech Communication: Fundamentals, Speech Perception Theory, and Technology* (Allyn and Bacon, Boston, 1999)
3. J.E. Cahn, Generating expression in synthesized speech. Master's thesis, MIT, 1989
4. J. Allen, M.S. Hunnicutt, D.H. Klatt, R.C. Armstrong, D.B. Pisoni, *From Text to Speech: The MITalk System* (Cambridge University Press, Cambridge, 1987)
5. G. Rosen, Dynamic analog speech synthesizer. J. Acoust. Soc. Am. **30**, 201–209 (1958)
6. P. Birkholz, Vocaltractlab [Online]. http://www.vocaltractlab.de
7. I. Steiner, Observations on the dynamic control of an articulatory synthesizer using speech production data. Ph.D. thesis, Saarland University, 2010
8. K. Iskaroust, L.M. Goldsteinta, D. Whalent, M.K. Tiedetb, P.E. Rubintc, CASY: the configurable articulatory synthesizer, in *Proceedings of International Congress of Phonetic Sciences* (2003), pp. 185–188
9. Z.-H. Ling, K. Richmond, J. Yamagishi, R.-H. Wang, Integrating articulatory features into HMM-based parametric speech synthesis. IEEE Trans. Audio Speech Lang. Process. **17**(6), 1171–1185 (2009)
10. M. Astrinaki, A. Moinet, J. Yamagishi, K. Richmond, Z.-H. Ling, S. King, T. Dutoit, Mage-HMM-based speech synthesis reactively controlled by the articulators, in *Proceedings of International Speech Communication Association Speech Synthesis Workshop (ISCA SSW8)* (2013), pp. 207–211
11. T. Dutoit, V. Pagel, N. Pierret, F. Bataille, O. van der Vrecken, The MBROLA project: towards a set of high quality speech synthesizers free of use for non-commercial purposes, in *Proceedings of International Conference on Spoken Language (ICSLP)* (1996), pp. 1393–1396
12. E. Moulines, F. Charpentier, Pitch-synchronous waveform processing techniques for text-to-speech synthesis using diphones. Speech Commun. **9**(5–6), 453–467 (1990)
13. H.T. Bunnell, D. Yarrington, K.E. Barner, Pitch control in diphone synthesis, in *Proceedings of ESCA/IEEE Workshop on Speech Synthesis* (1994), pp. 127–130
14. A.J. Hunt, A.W. Black, Unit selection in a concatenative speech synthesis system using a large speech database, in *Proceedings of IEEE International Conference on Acoustics, Speech and Signal Processing (ICASSP)* (1996), pp. 373–376
15. T. Raitio, H. Lu, J. Kane, A. Suni, M. Vainio, S. King, P. Alku, Voice source modelling using deep neural networks for statistical parametric speech synthesis, in *Proceedings of European Signal Processing Conference (EUSIPCO)* (2014), pp. 2290–2294
16. T. Raitio, A. Suni, L. Juvela, M. Vainio, P. Alku, Deep neural network based trainable voice source model for synthesis of speech with varying vocal effort, in *Proceedings of Interspeech* (2014), pp. 1969–1973
17. J. Yamagishi, T. Kobayashi, Y. Nakano, K. Ogata, J. Isogai, Analysis of speaker adaptation algorithms for HMM-based speech synthesis and a constrained SMAPLR adaptation algorithm. IEEE Trans. Audio Speech Lang. Process. **17**(1), 66–83 (2009)

18. T. Drugman, T. Raitio, Excitation modeling for HMM-based speech synthesis: breaking down the impact of periodic and aperiodic components, in *Proceedings of International Conference on Audio, Speech and Signal Processing (ICASSP)* (2014), pp. 260–264
19. H. Lu, Z.-H. Ling, M. Lei, C.-C. Wang, H.-H. Zhao, L.-H. Chen, Y. Hu, L.-R. Dai, R.-H. Wang, The USTC system for Blizzard challenge 2009, in *Proceedings of Blizzard Challenge Workshop* (2009)
20. L.-H. Chen, C.-Y. Yang, Z.-H. Ling, Y. Jiang, L.-R. Dai, Y. Hu, R.-H. Wang, The USTC system for Blizzard challenge 2011, in *Proceedings of Blizzard Challenge Workshop* (2011)
21. Y. Yu, F. Zhu, X. Li, Y. Liu, J. Zou, Y. Yang, G. Yang, Z. Fan, X. Wu, Overview of SHRC-Ginkgo speech synthesis system for Blizzard challenge 2013, in *Proceedings of Blizzard Challenge Workshop* (2013)
22. M. Plumpe, A. Acero, H. Hon, X. Huang, HMM-based smoothing for concatenative speech synthesis, in *Proceedings of International Conference on Spoken Language Processing (ICSLP)* (1998), pp. 2751–2754
23. J. Wouters, M. Macon, Unit fusion for concatenative speech synthesis, in *Proceedings of International Conference on Spoken Language Processing (ICSLP)* (2000), pp. 302–305
24. T. Okubo, R. Mochizuki, T. Kobayashi, Hybrid voice conversion of unit selection and generation using prosody dependent HMM. IEICE Trans. Inf. Syst. **E89-D**(11), 2775–2782 (2006)
25. V. Pollet, A. Breen, Synthesis by generation and concatenation of multiform segments, in *Proceedings of Interspeech* (2008), pp. 1825–1828
26. S. Tiomkin, D. Malah, S. Shechtman, Z. Kons, A hybrid text-to-speech system that combines concatenative and statistical synthesis units. IEEE Trans. Audio Speech Lang. Process. **19**(5), 1278–1288 (2011)
27. A. Sorin, S. Shechtman, V. Pollet, Refined inter-segment joining in multi-form speech synthesis, in *Proceedings of Interspeech* (2014), pp. 790–794
28. K.S.R. Murty, B. Yegnanarayana, Epoch extraction from speech signals. IEEE Trans. Audio Speech Lang. Process. **16**(8), 1602–1613 (2008)
29. H. Zen, T. Toda, M. Nakamura, K. Tokuda, Details of Nitech HMM-based speech synthesis system for the Blizzard Challenge 2005. IEICE Trans. Inf. Syst. **E90-D**(1), 325–333 (2007)
30. T. Raitio, A. Suni, J. Yamagishi, H. Pulakka, J. Nurminen, M. Vainio, P. Alku, HMM-based speech synthesis utilizing glottal inverse filtering. IEEE Trans. Audio Speech Lang. Process. **19**(1), 153–165 (2011)
31. T. Drugman, T. Dutoit, The deterministic plus stochastic model of the residual signal and its applications. IEEE Trans. Audio Speech Lang. Process. **20**(3):968–981 (2012)
32. H. Kawahara, H. Katayose, A. de Cheveigne, R. Patterson, Fixed point analysis of frequency to instantaneous frequency mapping for accurate estimation of F0 and periodicity, in *Proceedings of Eurospeech* (1999), pp. 2781–2784
33. R. Goldberg, L. Riek, *A Practical Handbook of Speech Coders* (CRC Press, Boca Raton, 2000)
34. B. Yegnanarayana, K.S.R. Murty, Event-based instantaneous fundamental frequency estimation from speech signals. IEEE Trans. Audio Speech Lang. Process. **17**(4), 614–624 (2009)

Chapter 2
Background and Literature Review

2.1 HMM-Based Speech Synthesis

HMM-based speech synthesis is one instance of statistical parametric speech synthesis, where hidden Markov models are used for statistical modeling. In HMM-based speech synthesis, the speech can be parameterized using different approaches such as harmonic plus noise model, sinusoidal model, and source-filter model. In this work, the speech is parameterized based on source-filter model (shown in Fig. 2.1). The most direct and simplest implementation of such a model uses an excitation signal (train of pulses or white noise) filtered with a linear time-invariant filter to produce the speech signal. The linear time-invariant filter models the spectral envelope of the glottal flow, vocal-tract resonances, and lip radiation effect. The excitation signal can be represented by voicing information and fundamental frequency, and the spectral envelope is represented by mel-cepstral coefficients [1]. The speech waveform can be reasonably reconstructed from the sequence of spectral and excitation parameters. In HMM-based speech synthesis, the spectral and excitation parameters are modeled by HMMs during training, and at the time of synthesis, these parameters are predicted from the HMMs for the given input text. The spectral and excitation parameters extracted from each frame form an observation vector. The observation vector includes both static and dynamic features. An example of an observation vector is shown in Fig. 2.2. Apart from mel-cepstral coefficients, various spectral representations, such as line-spectral frequencies (LSFs) [2], mel-generalized cepstral coefficients [3], and various excitation parameters (e.g., voicing strengths [4], aperiodicities [5]) can also be used.

K. S. Rao, N. P. Narendra, *Source Modeling Techniques for Quality Enhancement in Statistical Parametric Speech Synthesis*, SpringerBriefs in Speech Technology,
https://doi.org/10.1007/978-3-030-02759-9_2

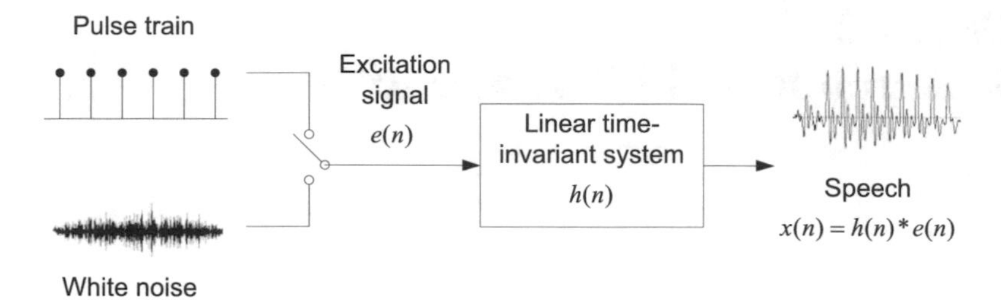

Fig. 2.1 Representation of human speech production using source-filter model

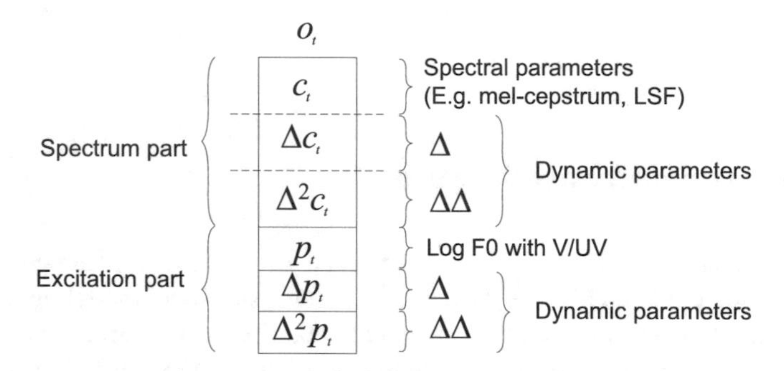

Fig. 2.2 An example of an observation vector

2.1.1 Hidden Markov Model

A hidden Markov model is a finite state machine which changes from state i to state j at each time step. Upon entering a state j at each time t, a continuous observation vector $\boldsymbol{o}_t$ is generated from the state output probability distribution $b_j(\boldsymbol{o}_t)$. For a state sequence of length T, $\boldsymbol{q} = \{q_1, q_2, \ldots, q_T\}$, the sequence of observations is defined as $\boldsymbol{O} = \{\boldsymbol{o}_1, \boldsymbol{o}_2, \ldots, \boldsymbol{o}_T\}$. An example of a three-state left-to-right HMM is shown in Fig. 2.3. An N-state HMM $\{\lambda\}$ is defined by a set of transition probabilities from state i to j, $\{a_{ij}\}_{i,j=1}^N$, state output probability distributions $\{b_i(.)\}_{i=1}^N$, and initial state probabilities $\{\pi_i\}_{i=1}^N$. The $\{b_i(.)\}_{i=1}^N$ are usually modeled by a single multivariate Gaussian distributions

$$b_i(\boldsymbol{o}_t) = \mathcal{N}(\boldsymbol{o}_t; \boldsymbol{\mu}_i, \boldsymbol{\Sigma}_i) \tag{2.1}$$

$$= \frac{1}{\sqrt{(2\pi)^d \mid \boldsymbol{\Sigma}_i \mid}} \exp\{-\frac{1}{2}(\boldsymbol{o}_t - \boldsymbol{\mu}_i)^T \boldsymbol{\Sigma}_i^{-1}(\boldsymbol{o}_t - \boldsymbol{\mu}_i)\} \tag{2.2}$$

Let $\boldsymbol{O} = [\boldsymbol{O}_1^{\mathrm{T}}, \boldsymbol{O}_2^{\mathrm{T}}, \ldots, \boldsymbol{O}_{\mathrm{T}}^{\mathrm{T}}]^{\mathrm{T}}$ and $\mathscr{W}$ be a set of speech parameters and the corresponding phoneme labels used for training the HMMs. The training of HMMs can be represented by the following equations:

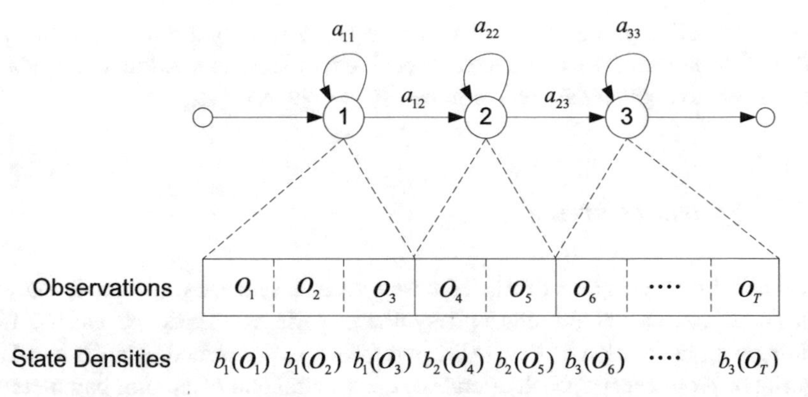

Fig. 2.3 A three-state left-to-right HMM with the illustration of an observation sequence and the state output probability distributions associated with each state

$$\textbf{Training: } \lambda_{\max} = \arg\max_{\lambda} p(\boldsymbol{O}|\lambda, \mathscr{W}) \tag{2.3}$$

$$p(\boldsymbol{O}|\lambda, \mathscr{W}) = \sum_{\forall \boldsymbol{q}} \pi_{q_0} \prod_{t=1}^{T} a_{q_{t-1}q_t} b_{q_t}(\boldsymbol{O}_t) \tag{2.4}$$

As both the model parameters and state sequences are unknown, it is difficult to estimate the HMM $\{\lambda\}$ which globally maximizes $p(\boldsymbol{O}|\lambda, \mathscr{W})$. However, the parameters of λ can be estimated by calculating the solution which maximizes $p(\boldsymbol{O}|\lambda, \mathscr{W})$ locally. Generally, the Baum-Welch algorithm [6], also called the expectation-maximization (EM) algorithm, is used to find this solution [7].

Let $\boldsymbol{o} = [\boldsymbol{o}_1^{\mathrm{T}}, \boldsymbol{o}_2^{\mathrm{T}}, \ldots, \boldsymbol{o}_{\mathrm{T}}^{\mathrm{T}}]^{\mathrm{T}}$ be the speech parameters generated at the time of synthesis for the given phoneme labels w. The synthesis from the HMMs can be represented by the following equations:

$$\textbf{Synthesis: } \boldsymbol{o}_{\max} = \arg\max_{\boldsymbol{o}} p(\boldsymbol{o}|\lambda_{\max}, w) = \arg\max_{\boldsymbol{o}} \sum_{\boldsymbol{q}} p(\boldsymbol{o}, \boldsymbol{q}|\lambda_{\max}, w) \tag{2.5}$$

The optimal speech parameter sequence is estimated by maximizing $p(\boldsymbol{o}, \boldsymbol{q}|\lambda_{\max}, w)$ with respect to $\boldsymbol{o}$ and $\boldsymbol{q}$. The optimal speech parameter sequence is approximated as follows:

$$\boldsymbol{o}_{\max} \approx \arg\max_{\boldsymbol{o},\boldsymbol{q}} p(\boldsymbol{o}|\boldsymbol{q}, \lambda_{\max}) P(\boldsymbol{q}|\lambda_{\max}, w) \approx \arg\max_{\boldsymbol{o}} p(\boldsymbol{o}|\boldsymbol{q}_{\max}, \lambda_{\max}) \tag{2.6}$$

where

$$\boldsymbol{q}_{\max} = \arg\max_{\boldsymbol{q}} P(\boldsymbol{q}|\lambda_{\max}, w) \tag{2.7}$$

The maximization problem of (2.7) can easily be solved by the state-duration probability distributions. The maximization problem of (2.6) is maximizing $p(\boldsymbol{o}|\boldsymbol{q}, \lambda,)$ with respect to $\boldsymbol{o}$ given the predetermined state sequence $\boldsymbol{q}_{\max}$.

2.1.2 System Overview

The basic block diagram of HMM-based speech synthesis system is shown in Fig. 2.4. It consists of training and synthesis parts as mentioned earlier. In the training part, estimation of the HMM parameters is performed using Baum-Welch algorithm. From every speech utterance, the spectral and excitation parameters are extracted. The spectral parameters such as mel-cepstral coefficients are modeled

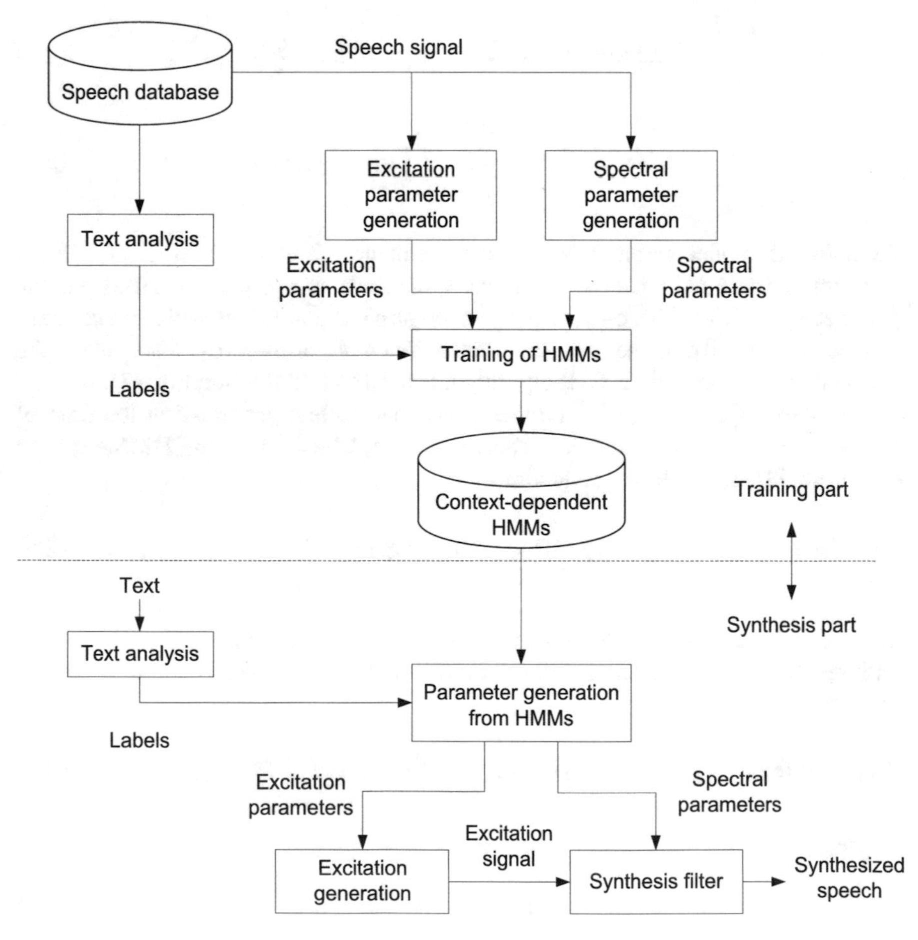

Fig. 2.4 The basic block diagram of HMM-based speech synthesis system (adapted from [8])

by continuous probability distribution (CD) HMMs. The excitation parameters which include F_0 consist of one-dimensional, continuous values that represent voiced regions and a discrete symbol that represents unvoiced regions. In order to simultaneously model both continuous values and discrete symbols, the conventional discrete or continuous HMMs cannot be used directly. To model such an observation sequence, a new kind of HMM based on multi-space probability distribution (MSD-HMM) has been used. MSD-HMM consists of a discrete HMM to model unvoiced symbol and a continuous mixture HMM to model continuous F_0 values. MSD-HMM assumes that F_0 values occur from one-dimensional space and unvoiced symbol occurs from zero-dimensional space. The spectral parameters and F_0 parameters are modeled simultaneously by separate streams in a multi-stream HMM [9].

2.1.3 Duration Modeling

Generally, in the case of standard HMMs, the state duration probability decreases exponentially with increase of duration, and as a result it is very simple to control the temporal structure of the speech parameter sequence. Instead, HMM-based speech synthesis typically uses a semi-Markov structure in which the state durations of each HMM are explicitly modeled by a multivariate Gaussian distribution [10]. The dimensionality of state duration density of an HMM is equal to the number of states in the HMM, and the nth dimension of state duration density corresponds to the nth state of the HMM.

2.1.4 Decision Tree-Based Context Clustering

Contextual features affect the spectrum, F_0, and duration of the speech utterance. To capture acoustic variations associated with the contextual features, context-dependent HMMs are used. As the number of contextual features increase, their combinations also increase exponentially. Therefore, the model parameters cannot be estimated accurately with a limited amount of training data. Furthermore, it is impractical to collect a speech corpus that includes all possible combinations of context factors with the sufficient number of examples. To overcome this problem, a decision tree-based context clustering technique is applied to the distributions of spectrum, F_0, and state durations. Similar states are clustered, and model parameters are tied among several context-dependent HMMs so that the estimation of model parameters can be performed robustly [11]. The state tying process is conducted in a hierarchical tree structure manner, and the tree size is automatically determined based on an information criterion called minimum description length (MDL) [12]. Since each of the spectrum, F_0, and duration has its own influential contextual factors, all three parameters are clustered independently. The state durations of each HMM are modeled by an n-dimensional Gaussian, and the context-dependent n-dimensional Gaussians are also clustered by a decision tree [13].

2.1.5 *Synthesis*

The synthesis part of the system is shown in the lower part of Fig. 2.4. First an input text to be synthesized is converted to context-dependent label sequence. According to the label sequence, a sentence HMM is constructed by concatenating context-dependent HMMs. The state durations of the sentence HMM are determined to maximize the output probability of state durations. Then a sequence of mel-cepstral coefficients and log F_0 values, including voiced/unvoiced decisions, are determined so as to maximize its output probability using the speech parameter generation algorithm [14]. The inclusion of dynamic features (delta and delta-delta) makes the generated speech parameters to be more realistic and smooth. Finally, the speech waveform is synthesized directly from the generated spectral and excitation parameters by using a synthesis filter, such as the mel-log spectral approximation filter [1] for mel-cepstral coefficients and all-pole filter for linear prediction-based spectral coefficients.

Due to the parametric representation and statistical modeling, the speech generated is smooth and intelligible but suffers from reduced naturalness compared to unit selection method. Several improvements have been introduced in HMM-based speech synthesis to improve the synthesis quality. One way of improving the quality of synthesis is through efficient parameterization and modeling of the excitation signal. Different issues addressed in this work include voicing detection and F_0 estimation, parametric and hybrid approaches of modeling excitation signal, and generation of creaky voice. Methods followed in literature for addressing each of these issues are discussed in the following section.

2.2 Voicing Detection and F_0 Estimation: A Review

One of the important source or excitation features is the pitch or F_0 with voicing/unvoicing detection. The pitch is a perceptual measure, and it is the frequency of pure tone that is matched by the listener to a more complex signal. F_0 is an objective measure, and it is computed as the fundamental frequency of speech signal. In the context of speech processing, pitch and F_0 are used interchangeably [15]. The voicing detection involves identification of the regions of speech with significant glottal activity. The voicing detection and pitch extraction are very much essential for speech coding and speech synthesis systems. Especially, the quality of HMM-based speech synthesis depends on the accuracy of voicing detection and

F_0 extraction. Errors in voicing detection and pitch estimation can contribute to unnaturalness and degradation in voice quality. In this section, we will review some of the important F_0 estimation methods.

The methods for estimating the F_0 from speech signal can be broadly categorized into three types: methods based on (1) time-domain characteristics, (2) frequency domain properties, and (3) both time and frequency domain properties. Some of the important works carried out in each of the methods are discussed below. One of the earlier F_0 estimation method is subharmonic summation method [16], which performs the pitch estimation based on the spectral compression model. The compression model is equivalent to the concept that each spectral component activates not only the elements of the central pitch processor but also those elements that have a harmonic relation with this component. The central pitch processor keeps track of the harmonics that give rise to the pitch. The contributions of the various subharmonic spectrum components add up, and the maximum of the subharmonic sum spectrum is the estimate of pitch. A pitch estimation method was proposed in [17] by refining the basic autocorrelation method. In this method, the autocorrelation of the speech segment is computed by dividing the autocorrelation of the windowed speech signal with the autocorrelation of the window. Among several pitch tracking methods, the method based on normalized autocorrelation is the most popular and simplest method [18]. For periodic signals, the autocorrelation function is also periodic. The first peak after the center peak in the autocorrelation sequence indicates the pitch period of the speech signal. A speech transformation and representation using adaptive interpolation of the weighted spectrum (STRAIGHT) method is used for extracting F_0 [19]. The STRAIGHT method extracts F_0 by using a wavelet-based instantaneous frequency analysis technique. At first, an instantaneous frequency based on estimate of F_0 is generated. Then, the best F_0 estimate is selected based on maximum carrier-to-noise ratio. In [20], the pitch is estimated by computing the cross correlation of the actual speech signal and synthesized voiced speech signals with different values of fundamental frequency. The position of the maximum peak in the cross correlation function indicates the estimated value of fundamental frequency. Yegnanarayana et al. [21] estimated the fundamental frequency by computing the reciprocal of successive epoch intervals obtained from zero frequency filtering (ZFF) method. The ZFF is one of the popular methods for extracting glottal closure instants from speech signal. In ZFF method, it is observed that the discontinuity due to impulse-like excitation is reflected across all frequencies including zero frequency and the resonances due to vocal-tract system are present at frequencies greater than 300 Hz. By designing a resonator at zero frequency, the information around zero frequency is greatly emphasized compared to the vocal-tract resonances. Hence, the speech signal is passed through a cascade of two resonators located at zero frequency, and the resulting signal is subtracted from its local mean. The positive zero crossings in the mean subtracted signal correspond to the epoch locations. The basic ZFF method uses a constant window size for local mean subtraction. Further enhancement of the ZFF method is performed by smoothing the estimated pitch by using the Hilbert envelope of the speech. The Hilbert envelope is used to compute the amplitude envelope of a

Table 2.1 Literature review on pitch estimation methods

Serial no.	Method	Description	Ref.
01	Subharmonic summation method (SHS)	Pitch estimation is performed based on the spectral compression model	[16]
02	Praat's autocorrelation method (AC)	Pitch is estimated using the modified autocorrelation which is computed by dividing the autocorrelation of the windowed speech signal with the autocorrelation of the window	[17]
03	Robust algorithm for pitch tracking (RAPT)	Pitch is estimated from the normalized autocorrelation of the speech	[18]
04	STRAIGHT	F_0 is estimated by using the wavelet-based instantaneous frequency analysis technique	[19]
05	Cross correlation method (CC)	Pitch is estimated by computing the cross correlation of the actual speech signal and synthesized voiced speech signals with different values of F_0	[20]
06	ZFF with uniform window (ZFFUW)	Pitch is extracted by passing the speech signal through the ZFF with a fixed window size equal to average pitch period of the utterance	[21]
07	ZFF with Hilbert envelope (ZFFHE)	Pitch estimated from the ZFFUW method is smoothed using the HE of the speech	[21]
08	Summation of residual harmonics (SRH)	F_0 is estimated from summation of harmonics of the residual signal computed for each frequency in the range $[F_{0,\text{min}}, F_{0,\text{max}}]$ of the speaker	[22]

signal. The Hilbert envelope is obtained from the magnitude of the analytic signal. An analytic signal is a complex signal, where the real part is the original signal (in this case, speech signal) and the imaginary part is the Hilbert transform of the original signal. A robust pitch estimation which performs efficiently in noisy and adverse conditions is proposed in [22]. In this method, the summation of harmonics of the residual signal is computed for each frequency in the range $[F_{0,\text{min}}, F_{0,\text{max}}]$ of the speaker. The frequency value having the maximum summation of harmonics is the estimated pitch value for a given frame. A brief summary of the literature on pitch estimation methods is presented in Table 2.1.

From the review of existing literatures [19–22] exploring time and frequency domain characteristics, it is observed that the performance of different pitch estimation methods is reasonably good in modal regions. But their performance degrades sharply in creaky or weak voiced regions. Hence, there is a need for a method which can efficiently perform voicing detection and F_0 estimation both in modal and creaky voiced regions.

In HMM-based speech synthesis, the pitch values estimated from one of the F_0 estimation methods are modeled by using HMMs. The existing HTS systems [23, 24] utilize the previously proposed pitch estimation methods for F_0 modeling. In literature, there are very few pitch estimation methods [25] which are specifically designed for HTS systems. Most of the current HTS systems [26–28] use

STRAIGHT [19] method for F_0 estimation. In [29], it has been observed that the STRAIGHT fails to extract correct F_0 values in creaky or vocal fry regions. They also observed that the vocal fry segments are wrongly classified as unvoiced. This resulted in the degradation of the synthesized speech quality. Robust algorithm for pitch tracking (RAPT) [18] which is based on the autocorrelation method is also used for F_0 estimation in HTS [30, 31]. In this method, the F_0 estimation errors such as F_0 doubling and halving can occur, and the decision regarding voicing tends to change at low frequency. In [32], the F_0 is estimated using a voting method which combines instantaneous frequency amplitude spectrum (IFAS) algorithm [33], STRAIGHT [19], and RAPT [18]. The voting method reduced the F_0 extraction errors such as F_0 halving/doubling and voiced/unvoiced swapping errors. Fujisaki's F_0 generation model [34] is used for modeling the F_0 contours in Mandarin HTS [35]. The F_0 generation model is a command-response model that describes F_0 contours on the logarithmic scale as the superposition of phrase and accent components (or tone components for tonal languages) [34].

Utilization of erroneous F_0 estimation method in HMM-based speech synthesis results in the development of erroneous F_0 models. During synthesis, due to erroneous F_0 models, the voiced (or unvoiced) frames may be wrongly classified as unvoiced (or voiced). Also, the generated F_0 values may not have the actual F_0 variations present in the natural speech. Hence, there is a need for an efficient F_0 estimation method which results in the improvement of the quality of HMM-based speech synthesis.

2.3 Source Modeling Approaches: A Review

Source modeling approaches proposed in the literature can be broadly categorized into parametric and hybrid approaches. In parametric approach, the excitation signal is represented in terms of number parameters and modeled using HMMs. During synthesis, the excitation signal is constructed from the parameters generated by HMMs. Some of important parametric methods of modeling the excitation signal are provided below. One of the initial parametric approaches for modeling the excitation was reported by Yoshimura et al. [4]. It consists of modeling the excitation parameters used in the mixed excitation linear prediction (MELP) [36] algorithm by hidden Markov models. During synthesis, the generated excitation parameters were used to construct the mixed excitation in the same way as in MELP algorithm. Later, Zen et al. used STRAIGHT [5]-based source model for generating the excitation signal [26]. They modeled F_0 and aperiodicity parameters by HMMs in order to enable the generation of excitation signal during synthesis stage [26]. In [37], the excitation signal is constructed by state-dependent filtering of pulse trains and white noise sequences. During training, the filters and pulse trains are jointly optimized through a procedure that resembles analysis-by-synthesis speech coding algorithms. In [38], the excitation signal is modeled using waveform interpolation (WI) technique. In the WI framework [39], each cycle of the excitation signal is represented by a characteristic waveform (CW). Each of the CWs is compactly represented by using principal component analysis and modeled by using HMMs. During synthesis, the excitation signal is constructed by using the CWs generated from the HMMs. In [40], Liljencrants-Fant (LF) model is utilized to generate the glottal source signal in HTS. The LF parameters are modeled by HMMs, and during synthesis, the generated LF parameters are used to control the glottal pulse shape. In [23], the excitation signal is constructed by modifying a single natural instance of glottal flow pulse according to the source parameters generated by HMMs. The glottal flow pulse is obtained by iterative adaptive glottal inverse filtering (IAIF) algorithm [41]. In [42], the excitation signal is generated by a combination of two segments. The first segment is only a small fraction of the real residual frame around glottal closure instant, and the second part is obtained from the model generated source parameters that represent the amplitude envelope and the energy of the residual waveform. In [43], the pitch-scaled spectrum is used to derive the periodic and aperiodic parts of the excitation signal. The periodic spectrum is compressed to reduce the dimensionality, and the aperiodic measure is fitted to a sigmoid function for integration into HTS. Table 2.2 briefs out some of the important and recent works on the parametric approach of source modeling.

Even though several parametric source modeling methods have been proposed in the literature [23, 40, 42, 43], there is still a necessity for an efficient source modeling method which can accurately model the excitation signal and results in the generation of speech signal close to natural quality. From the existing literature, it is observed that the excitation signal is modeled by either considering the time-domain or frequency-domain representation of the signal. Parameterizing the entire residual frame may tend to miss out some of the finer details in the excitation signal. Hence there is a need to identify the regions in the excitation signal which are perceptually important. Accurate modeling of these regions of excitation signal can result in the improvement of the quality of HMM-based speech synthesis. In this work, we are proposing a parametric source modeling method by accurately parameterizing and modeling the excitation signal.

The hybrid approach utilizes both parameters derived from statistical models and real excitation segments for generating the excitation signal. Compared to the parametric approach, the number of hybrid source modeling methods proposed in the literature is very less. In [44], a hybrid source modeling method is proposed where a codebook of pitch-synchronous residual frames is constructed during training. At the time of synthesis, the source signal is generated by selecting the suitable residual frames from the codebook based on target specification. Raitio et al. [45] proposed a unit selection method to select the appropriate glottal source pulses from the database based on target and concatenation costs. The glottal source pulses are extracted from the differentiated glottal volume velocity signal obtained from IAIF algorithm [41]. Drugman et al. proposed a hybrid approach based on deterministic plus stochastic model (DSM) [24]. The excitation signal is divided into two bands delimited by a maximum voiced frequency. The low-frequency band is considered as the deterministic component, and the high-frequency band is considered as the stochastic component. The deterministic component is the

Table 2.2 Literature review on parametric source modeling approach

Serial no.	Method	Description	Ref.
01	Mixed excitation	The excitation is modeled using pitch, band-pass voicing strengths, and Fourier magnitudes of first ten harmonics	[4]
02	STRAIGHT-based source model	The excitation is modeled using F_0 and aperiodicity parameters	[26]
03	Excitation model based on state-dependent filters	The excitation is modeled by state-dependent filters which are optimized through a procedure that resembles analysis-by-synthesis speech coding algorithms	[37]
04	Excitation model based on waveform interpolation (WI)	The excitation signal represented in terms of characteristic waveforms is modeled using principal component analysis	[38]
05	Excitation model based on Liljencrants-Fant (LF) model	The excitation signal is modeled using LF parameters, which control the shape of the glottal pulse	[40]
06	Excitation model based on glottal inverse filtering	The glottal flow pulse obtained from glottal inverse filtering is parameterized	[23]
07	Excitation model without voicing/unvoicing classification	The excitation signal of voiced/unvoiced region is modeled as a combination of two segments. The first segment is a small fraction of real residual frame around glottal closure instant, and the second part is parameterized in terms of the amplitude envelope and the energy of the residual waveform	[42]
08	Excitation model based on pitch-scaled spectrum	The pitch-scaled spectrum of the excitation signal is parameterized into periodic and aperiodic components	[43]

first eigenvector obtained by the principal component analysis of the residual frames. The stochastic component is obtained by modifying the spectral and amplitude envelopes of white Gaussian noise. Instead of using fixed maximum voiced frequency, DSM-based source model is enhanced by using time-varying maximum voiced frequency [46]. Some of the important works on the hybrid source modeling approach are provided in Table 2.3.

In the existing hybrid source models [44, 45], the phone-specific characteristics of the excitation signal are not thoroughly explored in modeling and generation of the source signal. In [24], certain preliminary studies have been carried out to investigate the phone dependency of the residual frames. Even though there are variations in the residual frames across different phonetic classes, it is grossly concluded that these variations are not significant. Our intuition is that the quality of the synthesized speech may be more natural if the phone-specific characteristics of the excitation signal are incorporated in the synthesized speech. In this work, the phone-dependent characteristics of the excitation signal are analyzed, and a hybrid source modeling method is proposed which is capable of generating the excitation signal specific to phones.

Table 2.3 Literature review on hybrid source modeling approach

Serial no.	Method	Description	Ref.
01	Excitation model using codebook	A codebook of residual frames is constructed. During synthesis, the appropriate residual frame is selected from the codebook based on target specification	[44]
02	Excitation model using glottal source pulses	Glottal source pulses are stored in the database. During synthesis, unit selection method is used to select the appropriate glottal source pulses from the database based on target and concatenation costs	[45]
03	Excitation model using deterministic plus stochastic model (DSM)	The excitation signal is divided into two bands delimited by a maximum voiced frequency. The low-frequency part or deterministic component is the first eigenvector obtained by the principal component analysis of the residual frames. The high-frequency part or the stochastic component is obtained by modifying the spectral and amplitude envelopes of the white Gaussian noise	[24]

2.4 Generation of Creaky Voice: A Review

Most of the existing source modeling approaches are capable of modeling and generating the speech with modal voice quality [23, 24, 26]. To produce good quality speech, modeling of creaky voice is important, as it is frequently produced by the speakers used for developing text-to-speech synthesis. Improper modeling of the creaky voice can lead to degradation in the quality of the synthesized speech. Methods used for modeling modal voice cannot be applied directly on creaky voice, as they exhibit dramatically different acoustic characteristics than that of modal voice. Two main issues involved in the generation of speech with creaky voice include creaky voice detection and source modeling of creaky voice. In this section, different methods followed in literature to address these two issues are discussed.

Creaky voice detection algorithm identifies the creaky voice regions in the speech utterance. In literature, there are very few approaches to automatically detect the creaky regions [47–49], though several methods exist for identifying the broader class, i.e., irregular phonation. In [47], an extension of the aperiodicity, periodicity, and pitch (APP) detector is proposed for automatic detection of irregular phonation including creaky voice. In the first step, the irregular frames are separated from the periodic frames by using the periodicity measure of the APP detector. In the second step, by using the "dip profile" of the average magnitude difference function (AMDF) in various frequency bands, the creaky voiced regions are identified. Ishi et al. [48] computed short-term power and intraframe periodicity (IFP) strength contours from the speech signal for differentiating the modal and creaky voiced regions. The interpulse similarity (IPS) measure is used to differentiate unvoiced and creaky regions. In [49], a creaky voice detection method is proposed by using two

Table 2.4 Literature review on creaky voice detection methods

Serial no.	Method	Description	Ref.
01	An extension of the aperiodicity, periodicity, and pitch (APP) detector	The "dip profile" of the average magnitude difference function (AMDF) in various frequency bands is used for detecting the creaky voiced regions	[47]
02	Ishi method	Short-term power and intraframe periodicity strength contours are computed from the speech signal for differentiating the modal and creaky voiced regions	[48]
03	Kane-Drugman (KD) method	Two new acoustic features are proposed for creaky voice detection. The first feature exploits the occurrence of secondary peaks in the linear prediction residuals of the creaky regions, and the second feature captures the strong impulse-like peak and long glottal pulse duration properties of the creaky regions	[49]

new acoustic features. The first feature exploits the occurrence of secondary peaks in the linear prediction (LP) residuals of the creaky regions, and the second feature captures the strong impulse-like peak and long glottal pulse duration properties of the creaky regions. Even though several methods try to identify the creaky regions, still there is a necessity for a method that can accurately detect creaky regions in the speech utterance. A brief summary of the existing creaky voice detection methods is provided in Table 2.4.

In literature, the existing methods perform creaky voice detection by exploring different properties of the creaky phonation. The existing creaky detection methods do not fully exploit the parameters related to glottal closure instants. As the occurrence of creakiness is a source characteristic, the complete exploration of parameters related to glottal closure instants can improve the performance of creak detection. In this work, the parameters related to the instants of significant excitation, also called as epoch parameters, are explored for identifying the creaky regions from the speech utterance.

For the generation of speech with creaky voice quality, efficient modeling of the creaky excitation signal is very much necessary. Very few researchers have attempted to develop a source model capable of producing the speech with creaky voice quality [25, 50, 51]. In [50], a source model capable of producing the creaky voice is developed by extending the deterministic plus stochastic model (DSM) of the residual signal [24]. In the extended version of the DSM, the deterministic and stochastic components are extracted separately from the open and closed periods of the creaky residual frames. Raitio et al. [25] synthesized the creaky voice by utilizing the GlottHMM F_0 tracker [23] and the extension of the DSM-based source model. In [51], two methods are proposed to generate the creaky voice. The first one is a rule-based method which applies pitch halving and amplitude scaling of the pitch-synchronous residual frames with random factors to generate the creaky

Table 2.5 Literature review on source modeling of creaky voice

Serial no.	Method	Description	Ref.
01	Hybrid source model using extension of DSM	The deterministic and stochastic components are extracted separately from the open and closed periods of the creaky residual frames	[50]
02	Hybrid source model of creaky voice	Two methods are proposed for modeling the creaky voice. The first method modifies the pitch and amplitude of the residual frames with random factors to generate the creaky excitation. The second method generates the creaky excitation by selecting appropriate creaky residual frames stored in the database	[51]

excitation. The second one is a data-driven approach where a database of creaky residual frames is developed, and during synthesis, the appropriate creaky residual frames are chosen from the database using the unit selection method. Table 2.5 provides some of the existing works on source modeling of creaky voice. In the previous work, the creaky excitation signal is generated using the previously stored instances of creaky residual frames and modifying the pitch and energy of residual frames with random values. A detailed analysis nature of creaky residual frames and analysis characteristics of primary and secondary excitations of creaky residual frames need to be performed. Creaky source model should be designed based on the analysis of nature of creaky residual frames. In this work, by utilizing the properties of creaky phonation, a hybrid source modeling method is proposed for modeling the creaky excitation signal.

2.5 Summary

In this chapter, initially, a brief overview of HMM-based speech synthesis system is provided. Then, different issues in source modeling, namely, voicing detection and F_0 estimation, parametric and hybrid source modeling approaches, and generation of creaky voice, are reviewed. From the existing voicing detection and F_0 estimation methods, it is observed that there is a need for a robust method which ensures accurate voicing detection and F0 estimation not only in modal voiced regions but also in creaky or low-voiced regions. On reviewing different parametric modeling methods, it is concluded that there is a need for a method which can accurately model the finer details in the excitation signal which are perceptually important. From review of existing hybrid source modeling methods, it is observed that there is a scope for proposing an efficient source model which can exploit the phone-specific and natural characteristics of the excitation. In HMM-based speech synthesis, the generation of creaky voice in addition to modal voice is important as it improves the naturalness of synthesized speech. The existing methods used for creaky detection do not fully exploit all possible parameters related to glottal closure instants. Hence a creaky voice detection needs to be proposed which can completely exploit all possible parameters related to glottal closure instants. The previous works on creaky source modeling do not completely explore the nature of creaky residual frames, and hence, there exists a scope for the design of creaky source modeling based on the analysis of characteristics of creaky residual frames.

References

1. T. Fukada, K. Tokuda, T. Kobayashi, S. Imai, An adaptive algorithm for mel-cepstral analysis of speech, in *Proceedings of the IEEE International Conference on Acoustics, Speech, and Signal Processing (ICASSP)* (1992), pp. 137–140
2. F. Itakura, Line spectrum representation of linear predictor coefficients of speech signals. J. Acoust. Soc. Am. **57**, S35–S35 (1975)
3. K. Tokuda, T. Kobayashi, T. Masuko, S. Imai, Mel-generalized cepstral analysis a unified approach to speech spectral estimation, in *Proceedings of the International Conference on Spoken Language Processing (ICSLP)* (1994), pp. 1043–1046
4. T. Yoshimura, K. Tokuda, T. Masuko, T. Kobayashi, T. Kitamura, Mixed-excitation for HMM-based speech synthesis, in *Proceedings of the Eurospeech* (2001), pp. 2259–2262
5. H. Kawahara, I. Masuda-Katsuse, A. de Cheveigne, Restructuring speech representations using a pitch-adaptive time-frequency smoothing and an instantaneous-frequency-based F0 extraction: possible role of a repetitive structure in sounds. Speech Commun. **27**(3–4), 187–207 (1999)
6. L.E. Baum, T. Petrie, G. Soules, N. Weiss, A maximization technique occurring in the statistical analysis of probabilistic functions of Markov chains. Ann. Math. Stat. **41**(1), 164–171 (1970)
7. L.R. Rabiner, A tutorial on hidden Markov models and selected applications in speech recognition. Proc. IEEE **77**(2), 257–286 (1989)
8. K. Tokuda, H. Zen, A.W. Black, HMM-based approach to multilingual speech synthesis, in *Text to Speech Synthesis: New Paradigms and Advances*, ed. by S. Narayanan, A. Alwan (Prentice-Hall, Upper Saddle River, 2004), pp. 135–153
9. S. Young, G. Evermann, M. Gales, T. Hain, D. Kershaw, X.-Y. Liu, G. Moore, J. Odell, D. Ollason, D. Povey, V. Valtchev, P. Woodland *The Hidden Markov Model Toolkit (HTK) Version 3.4* (2006). Available: http://htk.eng.cam.ac.uk/
10. H. Zen, K. Tokuda, T. Masuko, T. Kobayashi, T. Kitamura, Hidden semi-Markov model based speech synthesis system. *IEICE Trans. Inf. Syst.* **E90-D**(5), 825–834 (2007)
11. J.J. Odella, The use of context in large vocabulary speech recognition, Ph.D. dissertation, Cambridge University, 1995
12. K. Shinoda, T. Watanabe, MDL-based context-dependent subword modeling for speech recognition. J. Acoust. Soc. Jpn. (E) **21**(2), 79–86 (2000)
13. T. Yoshimura, K. Tokuda, T. Masuko, T. Kobayashi, T. Kitamura, Simultaneous modeling of spectrum, pitch and duration in HMM-based speech synthesis, in *Proceedings of the Eurospeech* (1999), pp. 2347–2350
14. K. Tokuda, T. Yoshimura, T. Masuko, T. Kobayashi, T. Kitamura, Speech parameter generation algorithms for HMM-based speech synthesis, in *Proceedings of the International Conference on Acoustics, Speech, and Signal Processing (ICASSP)* (2000), pp. 1315–1318
15. E.C. Zsiga, *The Sounds of Language: An Introduction to Phonetics and Phonology* (Wiley-Blackwell, Chichester, 2012)
16. D.J. Hermes, Measurement of pitch by subharmonic summation. J. Acoust. Soc. Am. **83**(1), 257–264 (1988)
17. P. Boersma, Accurate short-term analysis of fundamental frequency and the harmonics-to-noise ratio of a sampled sound. Inst. Phon. Sci. **17**, 97–110 (1993)

18. D. Talkin, A robust algorithm for pitch tracking (RAPT), in Speech Coding and Synthesis (Elsevier Science, Amsterdam, 1995), pp. 495–518
19. H. Kawahara, H. Katayose, A. de Cheveigne, R. Patterson, Fixed point analysis of frequency to instantaneous frequency mapping for accurate estimation of F0 and periodicity, in *Proceedings of the Eurospeech* (1999), pp. 2781–2784
20. R. Goldberg, L. Riek, *A Practical Handbook of Speech Coders* (CRC Press, Boca Raton, 2000)
21. B. Yegnanarayana, K.S.R. Murty, Event-based instantaneous fundamental frequency estimation from speech signals. IEEE Trans. Audio Speech Lang. Process. **17**(4), 614–624 (2009)
22. T. Drugman, A. Alwan, Joint robust voicing detection and pitch estimation based on residual harmonics, in *Proceedings of the Interspeech* (2011), pp. 1973–1976
23. T. Raitio, A. Suni, J. Yamagishi, H. Pulakka, J. Nurminen, M. Vainio, P. Alku, HMM-based speech synthesis utilizing glottal inverse filtering. IEEE Trans. Audio Speech Lang. Process. **19**(1), 153–165 (2011)
24. T. Drugman, T. Dutoit, The deterministic plus stochastic model of the residual signal and its applications. IEEE Trans. Audio Speech Lang. Process. **20**(3), 968–981 (2012)
25. T. Raitio, J. Kane, T. Drugman, C. Gobl, HMM-based synthesis of creaky voice, in *Proceedings of the Interspeech* (2013), pp. 2316–2320
26. H. Zen, T. Toda, M. Nakamura, K. Tokuda, Details of Nitech HMM-based speech synthesis system for the Blizzard Challenge 2005. IEICE Trans. Inf. Syst. **E90-D**(1), 325–333 (2007)
27. H. Zen, T. Toda, K. Tokuda, The Nitech-NAIST HMM-based speech synthesis system for the Blizzard Challenge 2006. IEICE Trans. Inf. Syst. **E91-D**(6), 1764–1773 (2008)
28. K. Oura, H. Zen, Y. Nankaku, A. Lee, K. Tokuda, A tied covariance technique for HMM-based speech synthesis. IEICE Trans. Inf. Syst. **E93-D**(3), 595–601 (2010)
29. H. Sil, E. Helander, J. Nurminen, M. Gabbouj, Parameterization of vocal fry in HMM-based speech synthesis, in *Proceedings of the Interspeech* (2009), pp. 1775–1778
30. HMM-based speech synthesis system (HTS). Available: http://hts.sp.nitech.ac.jp/
31. Q. Zhang, F. Soong, Y. Qian, Z. Yan, J. Pan, Y. Yan, Improved modeling for F0 generation and V/U decision in HMM-based TTS, in *Proceedings of the International Conference on Acoustics Speech and Signal Processing (ICASSP)* (2010), pp. 4606–4609
32. J. Yamagishi, Z. Ling, S. King, Robustness of HMM-based speech synthesis, in *Proceedings of the Interspeech* (2008), pp. 581–584
33. D. Arifianto, T. Tanaka, T. Masuko, T. Kobayashi, Robust F0 estimation of speech signal using harmonicity measure based on instantaneous frequency. IEICE Trans. Inf. Syst. **E87-D**(12), 2812–2820 (2004)
34. H. Fujisaki, K. Hirose, Analysis of voice fundamental frequency contours for declarative sentences of Japanese. J. Acoust. Soc. Jpn. (E) **5**(4), 233–242 (1984)
35. Q. Sun, K. Hirose, W. Gu, N. Minematsu, Generation of fundamental frequency contours for Mandarin speech synthesis based on tone nucleus model, in *Proceedings of the Interspeech* (2005), pp. 3265–3268
36. A. McCree, K. Truong, E. George, T. Barnwell, V. Viswanathan, A 2.4 kbit/s MELP coder candidate for the new U.S. Federal Standard," in *Proceedings of the International Conference on Acoustics, Speech and Signal Processing (ICASSP)* (1996), pp. 200–203
37. R. Maia, T. Toda, H. Zen, Y. Nankaku, K. Tokuda, An excitation model for HMM-based speech synthesis based on residual modeling, in *Proceedings of the International Speech Communication Association Speech Synthesis Workshop 6 (ISCA SW6)* (2007), pp. 131–136
38. J.S. Sung, D.H. Hong, K.H. Oh, N.S. Kim, Excitation modeling based on waveform interpolation for HMM-based speech synthesis, in *Proceedings of the Interspeech* (2010), pp. 813–816
39. W. Kleijn, Continuous representations in linear predictive coding, in *Proceedings of the International Conference on Acoustics, Speech and Signal Processing (ICASSP)* (1991), pp. 201–204
40. J. Cabral, S. Renals, J. Yamagishi, K. Richmond, HMM-based speech synthesiser using the LF-model of the glottal source, in *Proceedings of the IEEE International Conference on Acoustics, Speech and Signal Processing (ICASSP)* (2011), pp. 4704–4707

41. P. Alku, Glottal wave analysis with pitch synchronous iterative adaptive inverse filtering. Speech Commun. **11**(2–3), 109–118 (1992)
42. J.P. Cabral, Uniform concatenative excitation model for synthesising speech without voiced/unvoiced classification, in *Proceedings of the Interspeech* (2013), pp. 1082–1086
43. Z. Wen, J. Tao, S. Pan, Y. Wang, Pitch-scaled spectrum based excitation model for HMM-based speech synthesis. J. Signal Process. Syst. **74**(3), 423–435 (2013)
44. T. Drugman, A. Moinet, T. Dutoit, G. Wilfart, Using a pitch-synchrounous residual codebook for hybrid HMM/frame selection speech synthesis, in *Proceedings of the International Conference on Acoustics, Speech and Signal Processing, (ICASSP)* (2009), pp. 3793–3796
45. T. Raitio, A. Suni, H. Pulakka, M. Vainio, P. Alku, Utilizing glottal source pulse library for generating improved excitation signal for HMM-based speech synthesis, in *Proceedings of the International Conference on Acoustics, Speech and Signal Processing, (ICASSP)* (2011), pp. 4564–4567
46. T. Drugman, T. Raitio, Excitation modeling for HMM-based speech synthesis: breaking down the impact of periodic and aperiodic components, in *Proceedings of the International Conference on Audio, Speech and Signal Processing (ICASSP)* (2014), pp. 260–264
47. S. Vishnubhotla, C. Espy-Wilson, Automatic detection of irregular phonation in continuous speech, in *Proceedings of the Interspeech* (2006), pp. 949–952
48. C. Ishi, K. Sakakibara, H. Ishiguro, N. Hagita, A method for automatic detection of vocal fry. IEEE Trans. Audio Speech Lang. Process. **16**(1), 47–56 (2008)
49. J. Kane, T. Drugman, C. Gobl, Improved automatic detection of creak. Comput. Speech Lang. **27**(4), 1028–1047 (2013)
50. T. Drugman, J. Kane, C. Gobl, Modeling the creaky excitation for parametric speech synthesis, in *Proceedings of the Interspeech* (2012), pp. 1424–1427
51. T.G. Csapo, G. Nemeth, Modeling irregular voice in statistical parametric speech synthesis with residual codebook based excitation. IEEE J. Sel. Top. Signal Process. **8**(2), 209–220 (2014)

Chapter 3
Robust Voicing Detection and F_0 Estimation Method

3.1 F_0 Modeling and Generation in HTS

In HMM-based speech synthesis, the F_0 values extracted from every frame of speech are modeled by HMMs. At the time of synthesis, the F_0 values are generated from the HMMs by using maximum likelihood parameter generation (MLPG) algorithm [1]. MLPG algorithm generates the most likely sequence of observation vectors. From the given context label sequence, context-dependent HMMs are constructed. From the context-dependent HMMs, the distributions for the static, delta, and delta-delta parameters are obtained. From the given distributions of the static, delta, and delta-delta parameters, the algorithm determines the most likely sequence of generated parameters. The algorithm considers both static and dynamic (delta and delta-delta) parameters to preserve the natural variations. Using the generated F_0 values, the excitation signal is constructed. The fundamental frequency of speech has continuous values in voiced regions and undefined in unvoiced regions. During synthesis, first a discrete voicing/unvoicing decision is taken for each frame. Then, for every voiced frame, the F_0 value is generated. For unvoiced frame, no value is generated. To simultaneously model discrete voiced/unvoiced decision and continuous F_0 patterns, an HMM based on a multi-space probability distribution (MSD-HMM) [2] is used. The state output distribution of MSD-HMM is

$$b_\theta(o) = \begin{cases} c_v \mathcal{N}(o; \mu_\theta, \sigma_\theta), & o \epsilon \text{ voiced region} \\ c_{uv}, & o \epsilon \text{ unvoiced region} \end{cases} \tag{3.1}$$

where o is the observation at state θ, c_v and c_{uv} are the probabilities of voiced and unvoiced regions, μ_θ and σ_θ are the means and variances of Gaussian distribution of F_0 in voiced region. $b_\theta(o)$ represents the continuous probability density in voiced

K. S. Rao, N. P. Narendra, *Source Modeling Techniques for Quality Enhancement in Statistical Parametric Speech Synthesis*, SpringerBriefs in Speech Technology,
https://doi.org/10.1007/978-3-030-02759-9_3

regions and discrete probability in unvoiced region. At the time of synthesis, every HMM state is classified as voiced or unvoiced depending on whether c_v is greater than 0.5.

During training of F_0 patterns, the state posterior occupancy will be fully assigned as either voiced or unvoiced, depending on the voicing condition of the frame. Due to errors in pitch tracking, the voiced frames may be falsely marked as unvoiced. This kind of error mostly happens at creaky or low voicing regions. The pitch tracking methods typically may have F_0 halving and doubling errors. Due to F_0 halving and doubling errors, the pitch tracking methods may not capture the actual F_0 variations present in the speech utterance [3]. Building MSD-HMM using flawed F_0 values results in the development of improper models. At the time of synthesis, the voiced/unvoiced decision of each state is made independently, based on multi-space probability distribution of that state. Due to the development of erroneous models, the voiced states may be wrongly predicted as unvoiced. As a result of erroneous voicing decision, the synthesized speech sounds very dry or hoarse which leads to the degradation of the quality of the speech. Due to erroneous F_0 modeling, the F_0 values generated from the HMMs may not possess the actual F_0 variations present in the natural speech. This results in the synthesized speech to be monotonous and, subsequently, leads to the distortion of the original speaker characteristics. Hence, there is a necessity for a pitch tracking algorithm which can perform accurate voicing decision and pitch estimation at both strong and weak voicing regions.

3.2 Proposed Method for Voicing Detection and F_0 Estimation

Most of the existing voicing detection and F_0 estimation methods [4–6] extract F_0 values by detecting the periodicity of speech signal in either time or frequency domains. These methods work well in strong voicing regions. The performance will severely degrade, if the speech is weakly voiced, i.e., successive glottal cycles are not correlated. During the production of voiced speech, the vocal-tract system is excited by a sequence of impulse-like signals caused by the rapid closure of the glottis in each cycle. This impulse-like excitation is present in both strong and weak voicing regions. But, the speech signal waveforms vary significantly for different voicing regions due to time-varying nature of the vocal-tract system. In view of this, it is better to determine the locations of instants of impulse-like excitation and take the reciprocal of the interval between successive instants as the fundamental frequency. In this work, ZFF method is used to determine the instants of glottal closure [7]. Here, the size of window used in ZFF exploited for accurate voicing detection and F_0 estimation. By adaptively choosing appropriate window size, the strength of excitation for voiced speech is significantly higher compared with unvoiced speech. With suitable threshold on the strength of excitation, accurate voicing detection is performed. The smooth and accurate F_0 contour is extracted by frame-wise zero-frequency filtering of speech with appropriate window size.

3.2.1 *Zero-Frequency Filtering Method for Detecting the Instants of Significant Excitation*

ZFF method is based on the principle that the discontinuity due to impulse excitation is reflected across all frequencies including zero-frequency [7]. By designing a resonator at zero frequency, the information around zero frequency is greatly emphasized compared to the vocal tract resonances. As a result, the information regarding discontinuities due to impulse excitation can be easily obtained from the resulting zero-frequency resonator output. The system function of a zero-frequency resonator ($H(z)$) is given by

$$H(z) = \frac{1}{1 + a_1 z^{-1} + a_2 z^{-2}} \tag{3.2}$$

where $a_1 = -2$ and $a_2 = 1$. The above resonator de-emphasizes the characteristics of the vocal-tract system. A cascade of two such resonators, given by system function $G(z) = H(z)H(z)$, is used to significantly de-emphasize all the resonances of vocal tract system relative to zero frequency. Let $s[n]$ denote the input speech signal. The output of cascade of two resonators ($x_s[n]$) is given by

$$x_s[n] = s[n] * g[n]. \tag{3.3}$$

The output of the zero-frequency resonator $x_s[n]$ decays or grows as a polynomial function of time. Hence, it is difficult to directly detect the effect of discontinuities due to impulse excitation in the filtered output. The characteristics of discontinuities due to impulse excitation are extracted by computing the deviation of zero-frequency filtered output from the local mean. The window length used for computing the local mean is chosen to be around average pitch period. The resulting signal obtained after subtracting from the local mean is called zero-frequency filtered signal [7] ($y[n]$) and is given by

$$y[n] = x_s[n] - \frac{1}{2N+1} \sum_{m=-N}^{N} x_s[n+m] \tag{3.4}$$

where $2N + 1$ represents the length of window in terms of a number of samples. The time instants of negative to positive zero crossings of the zero-frequency filtered signal are called as the instants of significant excitation or glottal closure instants or epochs. The sequence of impulse locations derived from the zero-frequency filtered signal is used for estimating the fundamental frequency. Illustration of the zero-

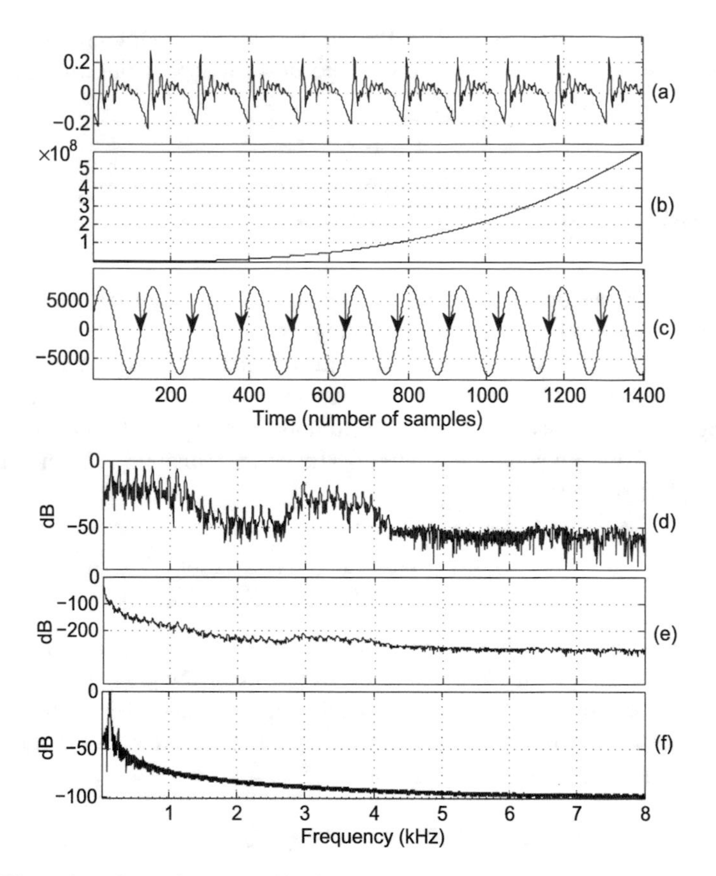

Fig. 3.1 Illustration of zero-frequency filtering method on a segment of speech signal. (**a**) Speech signal. (**b**) Zero-frequency resonator output. (**c**) Zero-frequency filtered signal. Log-magnitude spectrum of (**d**) speech signal, (**e**) zero-frequency resonator output, and (**f**) zero-frequency filtered signal

frequency filtering method on a segment of speech signal is provided in Fig. 3.1. The locations of epochs are marked by downward arrows in Fig. 3.1c.

3.2.2 Voicing Detection

Proposed method consists of two stages. First, each speech frame is classified as either voiced or unvoiced. Then for every voiced frame of speech, F_0 is estimated.

The voicing detection is performed based on the variation of the strength of excitation for different types of voicing regions. The strength of excitation is determined as the rate of closure of the vocal folds in each glottal cycle [8]. Sharper closure of the vocal folds corresponds to stronger excitation of the vocal-tract system. In voiced region, due to the vibration of vocal folds, the strength of excitation of vocal-tract system is significant. In unvoiced region, the vibration of vocal folds is completely absent, and hence the strength of excitation of vocal-tract system is very low. From zero-frequency filtered signal, the strength of excitation is computed as the slope of signal at epoch location [9]. The strength of excitation varies with the size of window used for local mean subtraction. Previous approaches [7, 9, 10] have not studied the effect of variation of window size on the strength of excitation as well as on the accuracy of pitch extraction. They have concluded that the window size equal to the average pitch period of the utterance is sufficient for local mean subtraction.

The strength of excitation plot computed for an utterance with a fixed window size equal to 2, 4, 6, 8, 10, 12, and 14 ms is shown in Fig. 3.2. The utterance shown in Fig. 3.2a consists of different voicing characteristics. The initial portion of the utterance corresponds to unvoiced region, the middle portion can be characterized as modal or strongly voiced region, and the end portion can be viewed as weakly voiced or creaky region. Figure 3.2b shows the strength of excitation at every epoch of an utterance for 2 ms window. In Fig. 3.2b, the epochs appear very close to each other. This is due to the occurrence of spurious epochs in between original epochs. The spurious epochs are falsely detected epochs by the ZFF method. The reason for occurrence of spurious epochs is explained in Sect. 3.2.3. As the window size

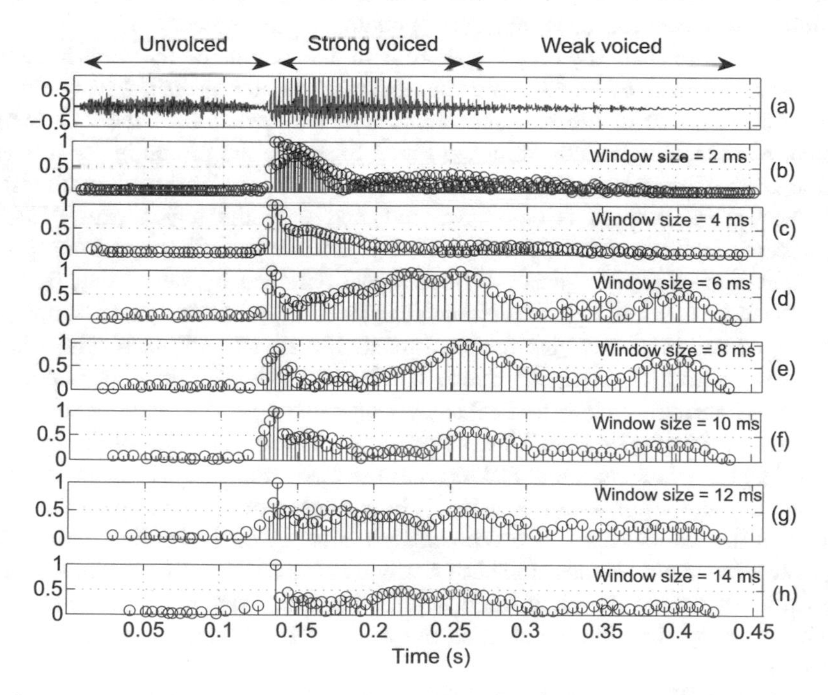

Fig. 3.2 (**a**) Speech signal. Strength of excitation with fixed window size of (**b**) 2 ms, (**c**) 4 ms, (**d**) 6 ms, (**e**) 8 ms, (**f**) 10 ms, (**g**) 12 ms, and (**h**) 14 ms

is increased, the spurious epochs are reduced. From Fig. 3.2b, it can be observed that the strength of excitation is high for modal voiced regions and low for both unvoiced and creaky voiced regions. If the speech is classified based on the strength of excitation obtained from the window size of 2 ms, then the weakly voiced regions are wrongly classified as unvoiced regions. Upon increasing the window size to 8 ms, it is observed that the strength of excitation in the creaky regions has been increased significantly, but the strength of excitation in the unvoiced region remains almost the same (refer Fig. 3.2e). This indicates that the strength of excitation of each voiced segment varies with the window size. The strength of excitation has a maximum value for a specific window size. For unvoiced regions, the strength of excitation does not vary much with the window size. This property of varying strength of excitation with window size is used for voicing detection.

3.2.3 Influence of Window Size on the Strength of Excitation

For demonstrating the influence of window size on the strength of excitation, the frequency domain illustrations of the sequence of steps in the ZFF method are exploited. The frequency domain interpretations for the sequence of time-domain operations of zero-frequency filtering of speech are shown in Fig. 3.1. Figure 3.1d shows the spectrum corresponding to the speech signal present in Fig. 3.1a. Figure 3.1e shows the spectrum of the output of the zero-frequency resonator, obtained by multiplying the spectrum of the speech signal with the frequency response of the resonator. Note that the spectrum of the ZFR output cannot be computed directly from the Fig. 3.1b due to the discontinuity of signal at the ends of segments [11]. In the Fig. 3.1e, it can be observed that most of the spectral information is de-emphasized, and only the region near 0 Hz has significant amplitude values. This operation is the low pass filtering of the speech. The spectrum of zero-frequency filtered signal obtained by subtracting from the local mean is shown in Fig. 3.1f. Finding the local mean of the signal is a low pass filtering operation. Subtracting the local mean of a signal with itself results in the high pass filtering of the signal. Successive low-pass (leading to ZFR output) and high-pass filtering (leading to ZFF signal) operations on the signal result in the band-pass filtering operation. This is evident from the observation of the spectrum of ZFF signal which shows a peak around the pitch frequency. In other words, it appears that the original signal in Fig. 3.1a is filtered by the band-pass filter. The choice of the window size affects the center frequency of the band-pass filter. The window size should be chosen in such a way that the center frequency of the band-pass filter is around the pitch frequency. If the center frequency is far from the pitch frequency, then instead of pitch frequency, other frequency components are enhanced.

Variation of the strength of excitation, ZFF signal and spectrum of ZFF signal for different window sizes are shown in Fig. 3.3. The top panel in each figure (Fig. 3.3a–g) shows the strength of excitation obtained from the ZFF signal. The

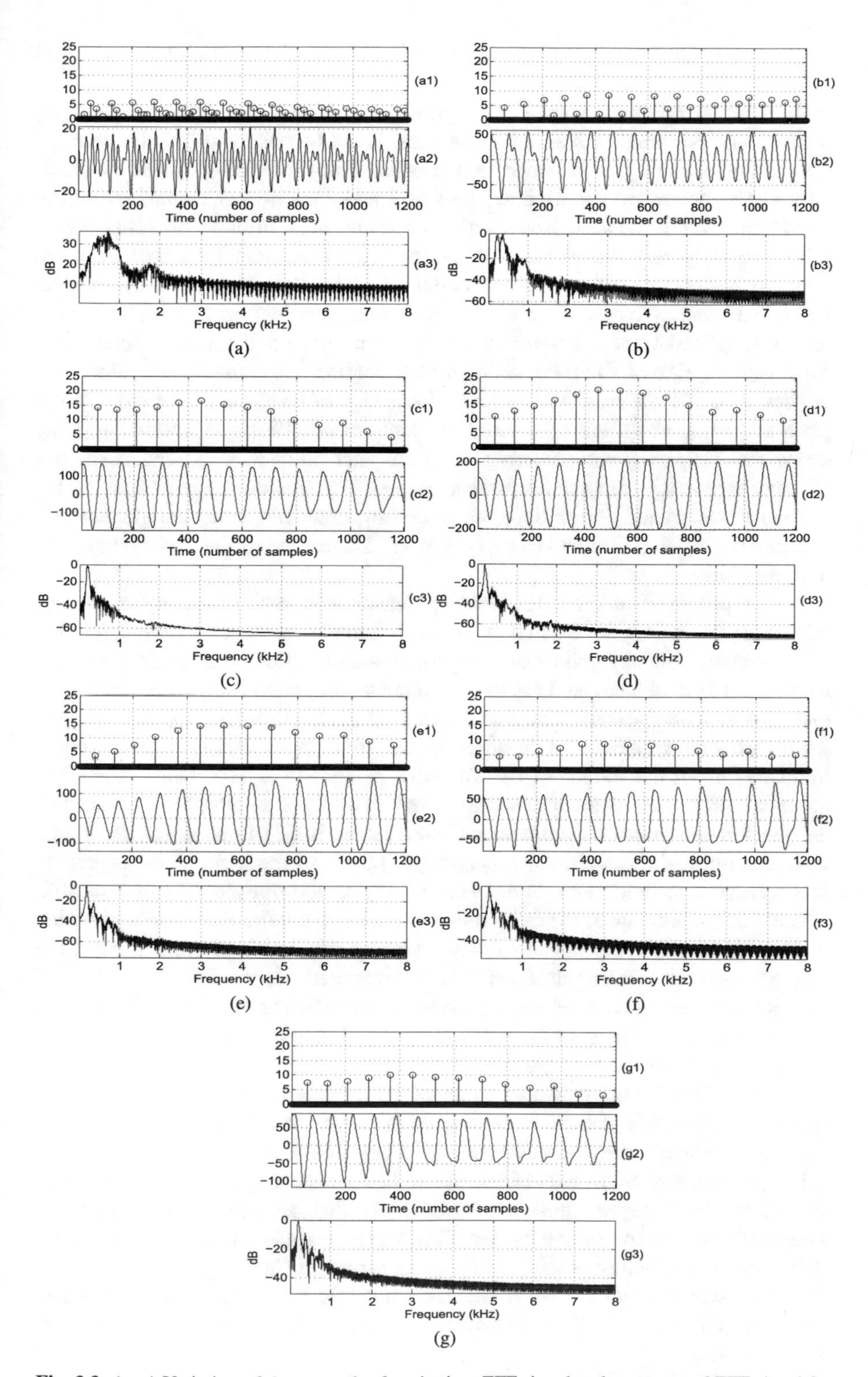

Fig. 3.3 (**a–g**) Variation of the strength of excitation, ZFF signal and spectrum of ZFF signal for different window sizes. (a1,b1,c1,d1,e1,f1,g1) Strength of excitation. (a2,b2,c2,d2,e2,f2,g2) ZFF signal. (a3,b3,c3,d3,e3,f3,g3) Log-magnitude spectrum. Window size of (**a**) 2 ms, (**b**) 4 ms, (**c**) 6 ms, (**d**) 8 ms, (**e**) 10 ms, (**f**) 12 ms, and (**g**) 14 ms

middle and the bottom panels of each figure show the ZFF signal and its spectrum, respectively. For window sizes of 2 and 4 ms, it can be observed that the spurious epochs are occurring in between true epochs. The center frequencies of band-pass filters formed by the window sizes of 2 and 4 ms do not coincide with the pitch frequency of the speech segment. As a result, in addition to pitch frequency, other frequency components are also present in the resulting ZFF signal. As more than one frequency components are present, the ZFF signal is deviating from the sinusoidal nature. Hence, there are spurious zero crossings in ZFF signal which results in the occurrence of spurious epochs. From Fig. 3.3, it can be observed that for a window size of 8 ms, the strength of excitation has a maximum value. The window size of 8 ms provides the center frequency of band-pass filter close to the pitch frequency of speech segment. The spectrum of ZFF signal (refer Fig. 3.3d) shows that frequency components around the pitch frequency are enhanced, and all other frequency components are suppressed. For window sizes of 2, 4, 6, 10, 12, and 14 ms, in addition to the pitch frequency, other frequency components are also present in the ZFF signal (see Fig. 3.3a–g). Hence, the strength of excitation is relatively low.

In the proposed method, the voicing detection is carried out by efficiently computing the strength of excitation. For each frame of speech, by varying the size of the window, the strength of excitation is computed. The window size is varied from 2 to 15 ms in steps of 1 ms to accommodate the pitch range (65–500 Hz) of male and female speakers. The maximum strength of excitation (MSE) obtained for a particular window size is stored as the strength excitation for that frame. Similarly for all the frames of speech, MSE is computed. The voicing detection is carried out by setting an appropriate threshold on the strength of excitation. The threshold is chosen for every utterance based on the MSE of that utterance. In this work, the optimal threshold value is found to be 8% of the MSE of the utterance. The optimal threshold value is derived by varying the threshold from 1 to 40% of MSE of the utterance, for the Keele database (details the database is provided in Sect. 3.2.5). For every threshold value, voicing decision is performed using the proposed method. Voicing decision error is computed which is the percentage of frames for which an error of voicing decision is made. The voicing decision error includes both miss rate and false alarm error rate. Voicing decision errors obtained for different values of the threshold are shown in Table 3.1. From the table, it is observed that for the threshold value of 8% of MSE, the voicing detection is optimum. Hence, 8% of MSE is considered as the threshold for voicing detection. For every database, the optimal threshold is selected which results in the lowest voicing decision error. In the context of synthesis, choosing appropriate threshold value has a direct impact on the quality of the synthesized speech. If the threshold value is not chosen appropriately, then it leads to increase in either miss rate or false alarm rate. Subsequently, voicing decision errors increase in the synthesized speech. Figure 3.4 shows the speech signal, MSE obtained for every speech frame, window size corresponding to MSE and detected voiced part of the speech signal (shown in bounding box).

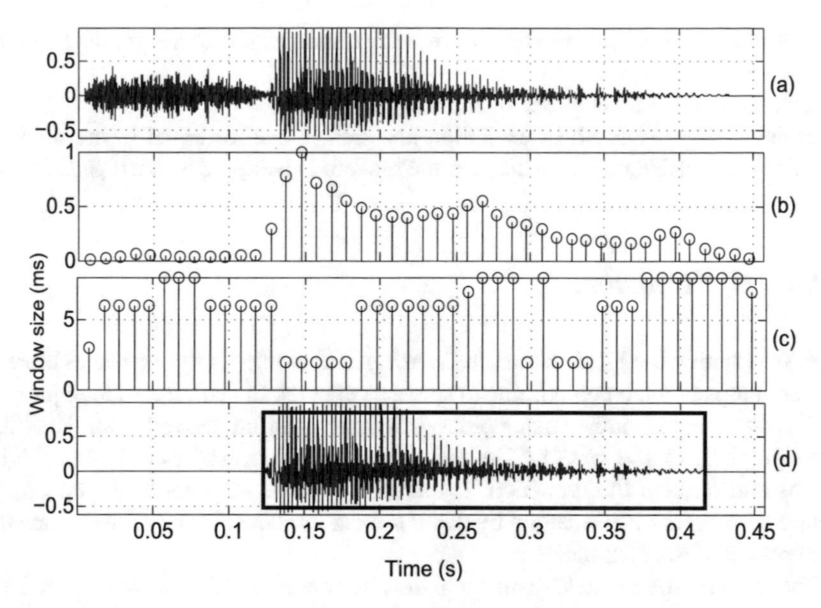

Fig. 3.4 (**a**) Speech signal. (**b**) MSE obtained for every speech frame. (**c**) Window size corresponding to MSE. (**d**) Voiced part of the speech signal (shown in bounding box)

Table 3.1 Voicing decision errors for different threshold values

Threshold values (%)	Voicing decision errors (%)
1	6.83
4	4.81
6	3.89
8	2.49
10	3.19
12	3.68
16	4.12
20	5.76
24	8.10
28	10.98
32	13.25
36	15.12
40	16.34

The sequence of steps for voicing detection is given below.

1. Pass the speech signal through the zero-frequency filter.
2. By varying the window size from 2 to 15 ms, compute the strength of excitation from the zero-frequency filtered signal.
3. For every window size, compute the sum of the strength of excitation of epochs present in the frame.

4. Determine the maximum strength of excitation and the corresponding window size for each frame.
5. Threshold value is set as 8% of MSE of the utterance.
6. Frames whose strength of excitation are greater than or equal to the threshold value are considered as voiced, and the rest are considered as unvoiced.

3.2.4 F_0 Estimation

While estimating the F_0 from the speech using ZFF method, at some places there are sudden jumps in pitch contour due to spurious epochs. In [10], the sudden jumps in pitch contour are smoothed using pitch period information derived from the Hilbert envelope (HE) of the speech. The HE of the speech enhances only the original epochs and ensures the reduction in spurious epochs. In general, the epochs are computed from the ZFF method by using a fixed window size equal to the average pitch period of the utterance.

The main reason for sudden jumps in the pitch contour is due to the occurrence of spurious epochs between original epochs. If the ZFF signal contains only one pitch frequency component, then the ZFF signal is almost a perfect sinusoid signal. If the ZFF signal contains other frequency components in addition to pitch frequency component, then the ZFF signal deviates from the sinusoidal nature. It is observed from many examples that the ZFF signal deviating from the sinusoidal nature contains more positive zero crossings (leading to spurious epochs), which leads sudden jumps in pitch contour. The reason for deviation from the sinusoidal nature of the ZFF signal is due to the selection of improper window length for the mean subtraction. This fact can be illustrated with an example in Fig. 3.5, which shows the speech signal and different ZFF signals obtained for the window lengths of 4, 6, and 8 ms. For the window length of 4 ms, the initial portion of the ZFF signal is a pure sinusoid, and as we move toward the end, the sinusoidal nature of the ZFF signal decreases, and the spurious epochs can be observed in this region. For the window length of 6 ms, the middle portion of the ZFF signal is a perfect sinusoid, and the end portion is still deviating from the sinusoidal nature. As the window length is increased to 8 ms, the end portion of the ZFF signal is close to the sinusoidal nature, and the initial portion is deviating from the sinusoidal behavior. In this case, the spurious epochs in the end portion are completely reduced, and the spurious epochs can be observed in the initial portions of the ZFF signal. From these observations, it can be inferred that if a uniform window length is used for zero-frequency filtering, then it can lead to the generation of spurious epochs during some regions in the utterance.

Figure 3.6 shows speech utterance, F_0 contours obtained from the window size of 3 and 6 ms and original F_0 contour. Sudden jump of F_0 contour at around 1.5 s can be observed in Fig. 3.6b for a window size of 3 ms, and it is reduced by changing the window size to 6 ms (Fig. 3.6c). But in Fig. 3.6c, the sudden jump in the F_0 contour is observed at around 1.2 s. This indicates that a uniform window size can produce

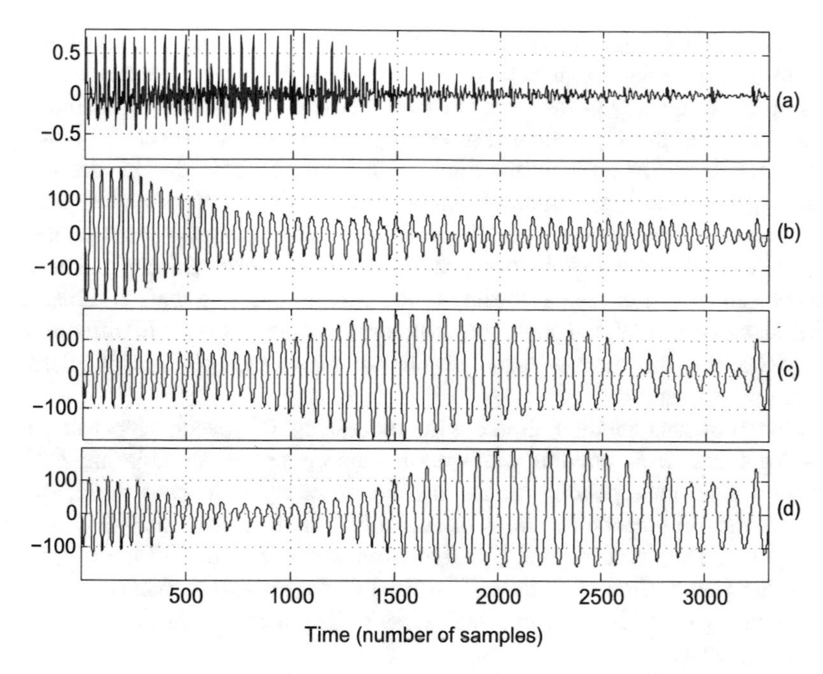

Fig. 3.5 (**a**) Speech signal. Zero-frequency filtered signal with fixed window size of (**b**) 4 ms, (**c**) 6 ms, and (**d**) 8 ms

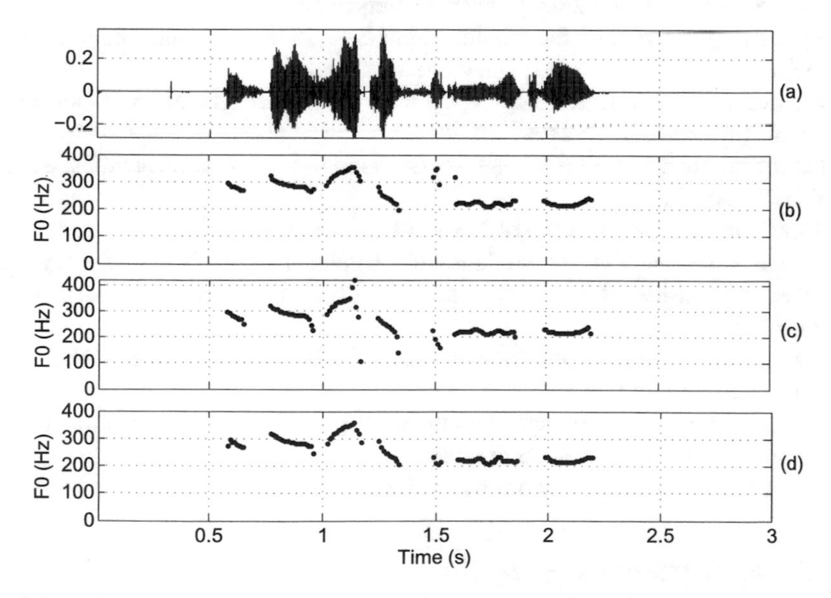

Fig. 3.6 (**a**) Speech signal. F_0 contours obtained from window size of (**b**) 3 ms and (**c**) 6 ms. (**d**) Original F_0 contour indicating the ground truth values given by electroglottograph (EGG) signals

smooth F_0 contour only at certain regions of the speech whose pitch frequency is close to the center frequency of the band-pass filter. For the speech regions whose pitch frequency is far from the center frequency, there are possibilities of sudden jumps in the F_0 contour. Smooth F_0 contour can be ensured by using an optimal window size suitable for the speech segment. This can be achieved by processing the speech segments of short duration using ZFF instead of using the entire speech utterance at a time. The main reason for using the short segments is that the variability of the pitch during short segments is very unlikely, and hence the window size will be chosen according to the pitch in that segment. Whereas, if the entire utterance is used, the pitch may fluctuate at different places in different ways, and by choosing the window size equivalent to the average pitch leads to errors in F_0 contour estimation.

In the proposed method, frame-wise processing of speech is performed for obtaining accurate F_0 contour. The speech frame is passed through the ZFF with a window size corresponding to the maximum strength of excitation. From the output of the ZFF, epoch locations are identified, and subsequently, average pitch is computed. The average pitch obtained from the output of ZFF is the estimated F_0 for that frame. Similarly, the F_0 is estimated for all speech frames. Processing the speech signal with an appropriate window length results in the generation of accurate F_0 value.

The sequence of steps for F_0 estimation is given below.

1. Pass the speech signal through the zero-frequency filter.
2. By varying the size of the window from 2 to 15 ms, compute the strength of excitation from the zero-frequency filtered signal.
3. For every window size, compute the sum of the strength of excitation of the epochs present in that frame.
4. Determine the maximum strength of excitation and the corresponding window size for each frame.
5. Threshold value is set as 8% of MSE of the utterance.
6. Frames whose strength of excitation are greater than or equal to the threshold value are considered as voiced, and the rest of the frames are considered as unvoiced.
7. The speech frame is passed through the ZFF with a window size corresponding to the MSE of that frame.
8. F_0 of the frame is determined from the output of ZFF by computing the mean of the epoch intervals present in the frame.
9. Similarly, the F_0 is estimated for all voiced frames.

3.2.5 *Performance Evaluation*

We intent to use the proposed method in HMM-based speech synthesis, which requires accurate F_0 value from the clean speech. Hence, the performance evaluation of the proposed voicing detection and pitch estimation method is carried out on the clean speech data. The performance of the proposed method is assessed using the following three measures.

- **Voicing Decision Error (VDE)** is the percentage of frames for which an error of voicing decision is made.
- **Gross pitch error (GPE)** is the proportion of voiced frames for which the estimated F_0 value deviates from the reference value by more than 20%.
- **Fine pitch error (FPE)** is the standard deviation of the absolute value of the difference between the estimated and reference pitch values. Here the finer variation of estimated pitch from the reference pitch needs to be computed. Hence, only those frames whose error is below a threshold of 20% are considered.

The proposed method is compared with eight other existing pitch tracking methods.

1. **Get_F0**: Available in Entropic signal processing system (ESPS) package, this method is an implementation of the RAPT method [4].
2. **Praat's Autocorrelation method (AC)**: This algorithm estimates F_0 on the basis of an accurate autocorrelation method [5].
3. **Cross-correlation method (CC)**: This method utilizes the cross-correlation function for extracting the pitch [6].
4. **Subharmonic summation method (SHS)**: This technique performs the pitch estimation based on the spectral compression model [12].
5. **STRAIGHT**: This method is based on the fixed point analysis of the speech [13].
6. **Summation of residual harmonics (SRH)**: This technique estimates the pitch based on the harmonicity of the residual signal [14].
7. **ZFF with uniform window size (ZFFUW)**: In this method, the pitch is extracted by passing the speech signal through the ZFF with a fixed window size equal to the average pitch period of the utterance [10].
8. **ZFF with Hilbert envelope (ZFFHE)**: The pitch estimated from the ZFFUW method is smoothed by using the HE of the speech [10].

Since ZFFUW and ZFFHE provide only F_0 estimates, the voicing detection for these methods are performed by choosing a threshold on the strength of excitation. All methods were used with their default parameter values for which they were optimized. The pitch values are derived by considering a frame size of 25 ms and a frame shift of 10 ms. The range of F_0 is set to [65 Hz, 500 Hz].

The performance of the existing and the proposed pitch estimation algorithms are evaluated on Keele [15] and Centre for Speech Technology Research (CSTR) databases [16]. These two databases are specially designed for the evaluation of the pitch tracking algorithms. The Keele database consists of speech utterances from five male and five female speakers. Each speaker is uttering a short story of duration 30 s. The CSTR database contains 5 min of speech, each read from one adult male and one adult female. The speech signal sampled at 16 kHz was considered for evaluation, and the reference pitch values provided in the database are used as the ground truth. In Sect. 3.2.3, the procedure for computing the optimal threshold value

Table 3.2 Performance of pitch tracking methods

Method	Keele database VDE (%)	GPE (%)	FPE	CSTR database VDE (%)	GPE (%)	FPE
Get_F0	9.28	9.88	4.04	7.58	8.48	6.12
Praat's autocorrelation method (AC)	12.35	5.34	3.69	9.43	5.23	6.28
Crosscorrelation method (CC)	11.26	6.89	3.37	6.27	6.81	6.73
Subharmonic summation method (SHS)	9.12	10.77	2.39	7.92	8.93	5.86
Speech transformation and representation using adaptive Interpolation of weiGHTed spectrum method (STRAIGHT)	8.20	3.77	3.97	8.61	4.98	6.10
Summation of residual harmonics method (SRH)	8.92	3.83	4.62	7.33	4.17	6.28
ZFF with uniform window size (ZFFUW)	6.14	2.93	4.55	6.92	3.39	6.97
ZFF with Hilbert envelope (ZFFHE)	5.28	2.60	4.47	5.65	1.94	6.88
Proposed method	2.49	2.46	4.38	3.10	1.62	6.49

for the Keele database is explained. The similar procedure is followed for computing the optimal threshold value for the CSTR database.

VDE, GPE, and FPE for different pitch tracking methods are given in Table 3.2. VDE and GPE are significantly lower for the proposed method, compared to other methods. The main reason for the lowest VDE by the proposed method on both databases is due to the use of the strength of excitation for voicing detection, instead of periodicity or energy of the signal. The voicing decision performed based on the strength of excitation is more accurate compared to other methods. The reason for reduced GPE is that, the frame-wise zero-frequency filtering of speech results in smooth and accurate F_0 contours. In the case of other pitch tracking methods, there were F_0 extraction errors such as F_0 halving or doubling. FPE is little higher compared with other methods. This might be explained by the fact that the proposed method reduces the GPE significantly. Low GPE indicates that the number frames with pitch frequency less than 20% of the reference value is large. As large number of frames are considered, the deviation in absolute F_0 error is slightly more. Results clearly indicate the effectiveness of the proposed method over other pitch tracking methods.

From the evaluation results, it can be observed that the proposed method demonstrated significant improvement in voicing detection. In particular, even in creaky regions, the accuracy of voicing detection is very high by the proposed method. The significance of the proposed method in identifying the creaky regions as voiced is illustrated by comparing with two other methods, namely, RAPT and STRAIGHT. Figure 3.7a shows the speech signal with creaky regions obtained from a male speaker of Keele database. Figure 3.7b–d shows the voiced regions detected

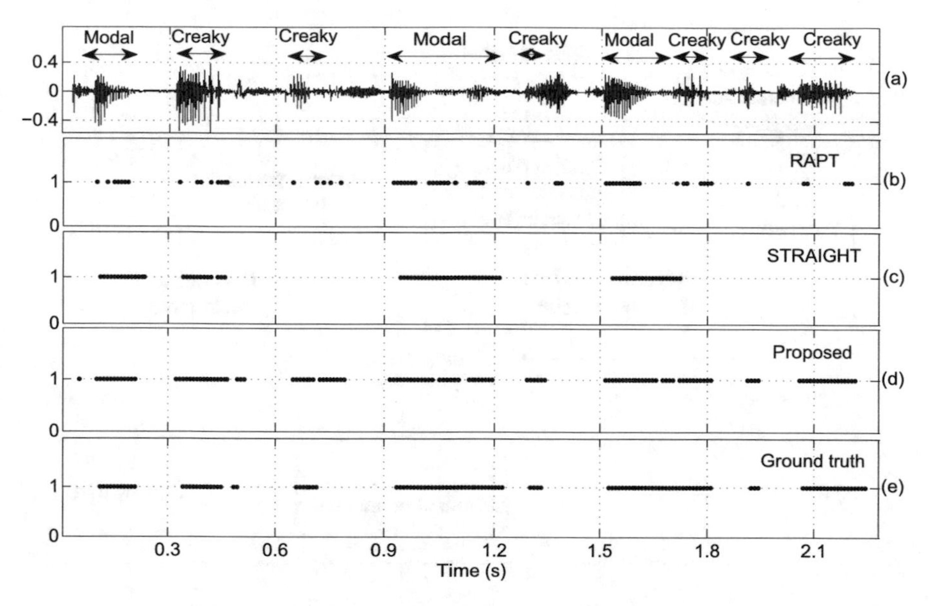

Fig. 3.7 (**a**) Speech signal containing both modal and creaky voiced regions. Voiced regions detected from (**b**) RAPT, (**c**) STRAIGHT, and (**d**) proposed method. (**e**) Original voiced regions indicating the ground truth values obtained from EGG signals. Voiced and unvoiced regions are marked as dots and empty spaces, respectively

from RAPT, STRAIGHT, and proposed method, respectively. The original voiced regions obtained from EGG signals are shown in Fig. 3.7e. In these figures, voiced and unvoiced regions are marked as dots and empty spaces, respectively. From the detected voiced regions in Fig. 3.7b, c, it can be observed that the RAPT and the STRAIGHT methods detect creaky regions as unvoiced. In the RAPT method, some of the modal voiced regions are also marked as unvoiced. But in Fig. 3.7d, it can be observed that in addition to modal voiced regions, the proposed method identified the creaky regions as voiced.

3.3 Implementation of the Proposed Voicing Detection and F_0 Extraction in HTS Framework

The overall block diagram of HMM-based synthesis system including proposed voicing detection and F_0 extraction method is shown in Fig. 3.8. The HMM-based speech synthesizer is implemented using publicly available HTS toolkit [17]. The spectrum part consists of 34-th order Mel-cepstral coefficients ($Fs = 16\,\text{kHz}$) and their delta and delta-delta coefficients. The spectrum part is modeled by CD-HMMs. The excitation part consists of log-fundamental frequency (log F_0) and its delta and delta-delta coefficients. The fundamental frequency patterns are modeled by

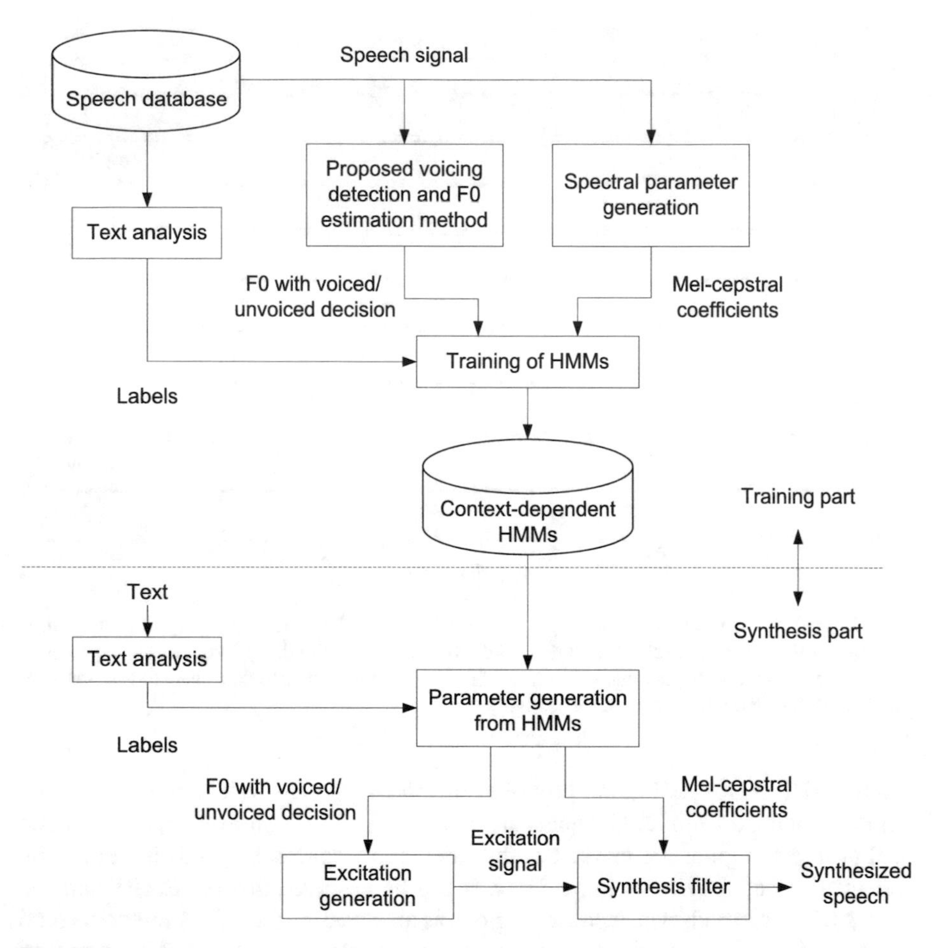

Fig. 3.8 HMM-based speech synthesis system with the proposed voicing detection and F_0 estimation method

MSD-HMMs. The output probabilities of spectrum and excitation part are modeled using a single Gaussian distribution with diagonal covariance. Both CD-HMMs and MSD-HMMs have five emitting states. From each state, the transition is allowed only to the next immediate state. The temporal structure of speech is modeled by the state-duration densities of HMMs. The state durations of each phoneme HMM are modeled by using a single Gaussian distribution with diagonal covariance. The Mel-cepstral coefficients and F_0 values are extracted from a frame size of 25 ms with a frame shift of 5 ms.

For the given input text, the text analysis module provides a sequence of phones along with its positional and contextual features. Using the phonetic labels having time alignment information, monophone HMMs are trained using the segmental K-means and expectation-maximization algorithm. The monophone HMMs are converted into context-dependent HMMs, and the model parameters are reestimated again. Decision tree-based context clustering technique [18, 19] is applied to the context-dependent HMMs. The question set consists of a standard list of 53 positional and contextual features provided in the basic HTS toolkit [17]. At each leaf node of the decision tree, the model parameters are tied and reestimated again. In the proposed system, both Mel-cepstral coefficients and F_0 streams were considered during the alignment step of reestimation.

At the time of synthesis, the input text to be synthesized is converted to a context-based label sequence. According to the label sequence, a sentence HMM is constructed by concatenating context-dependent HMMs. The state durations of the sentence HMM are determined to maximize the output probability of the state durations. Then a sequence of Mel-cepstral coefficients and log F_0 values, including voiced/unvoiced decisions, is determined by maximizing the output probability [1]. The generated Mel-cepstral coefficients and log F_0 values are often excessively smoothed due to statistical averaging. Global Variance technique is used to alleviate the over-smoothing effect [20]. Simple pulse/noise excitation is used to generate the excitation signal. Simple pulse/noise excitation scheme generates a sequence of pulses for voiced speech and white noise for unvoiced speech. The pulses are positioned according to the generated pitch period. The generated excitation signal is given as input to Mel-log spectral approximation (MLSA) filter to synthesize the speech.

3.4 Evaluation

The proposed method is evaluated using four English speakers from CMU Arctic speech database [21]. Out of four, two are females (SLT and CLB), and other two are male (AWB and KSP) speakers. For each of the speaker, the training set consists of about 1100 phonetically balanced English utterances. The duration of the training set is about 56, 64, 79, and 59 min for SLT, CLB, AWB, and KSP speakers, respectively. For every speech utterance, the corresponding phonetic transcriptions are available in the CMU Arctic speech database.

The proposed voicing detection and F_0 estimation method is compared with two other F_0 estimation methods, namely, STRAIGHT [13] and RAPT [4]. STRAIGHT method extracts F0 using the wavelet-based instantaneous frequency analysis technique. RAPT method is based on normalized autocorrelation function. The time interval between the first peak and the center peak in the autocorrelation sequence indicates the pitch period of the speech signal. The main reason for selecting these two methods for comparison is that these methods are widely used for F_0 estimation in HTS [22–25]. Three HMM-based speech synthesis systems are developed using the F_0 values estimated from three methods, namely, proposed, STRAIGHT, and RAPT. Except F_0 extraction step, all other steps in training and synthesis remain the same for three HTS systems. Twenty sentences which were not part of training data were used for the evaluation purpose. The evaluation of proposed method is performed in two steps. In the first step, evaluation of the voicing detection is performed and in the second step, subjective evaluation of the overall quality of the synthesized speech is carried out.

3.4.1 Evaluation of Voicing Detection

During synthesis, depending on the voicing probability c_v (as explained in Sect. 3.1), every state is classified as either voiced or unvoiced. Suppose a state which should be voiced (or unvoiced) may be wrongly classified as unvoiced (or voiced) due to erroneous voicing probability c_v. This happens due to wrong voicing detection by the pitch estimation method which results in the development of erroneous F_0 models. In order to check whether the proposed method has reduced the voicing/unvoicing errors significantly, the evaluation of the voicing detection is carried out.

The evaluation of the voicing detection is carried out by checking out how many of the voiced (or unvoiced) states in the synthesized speech are wrongly classified as unvoiced (or voiced). Wrong voicing/unvoicing decision results in the usage of inappropriate excitation signals for the synthesis of the speech. Usually, two types of voicing/unvoicing decision errors may occur, namely, (1) voiced states wrongly classified as unvoiced (v⇒u errors) and (2) unvoiced states wrongly classified as voiced (u⇒v errors). From the perceptual observation of the synthesized speech, it is noticed that v⇒u errors contributes more to the perceptual degradation of the voiced quality compared with u⇒v errors. v⇒u and u⇒v errors are determined for the HTS systems developed using F_0 estimated from RAPT, STRAIGHT, and proposed methods. Voicing/unvoicing condition of every state in the synthesized speech is manually compared with the natural utterance. The voicing/unvoicing decision error is computed as the ratio of total number of states having wrong voicing/unvoicing decision to total number of states present in the test set of 20 sentences (total number of states = total number of phones present in 20 sentences × 5). For the test set considered for evaluation, manual annotated phonetic labels and voicing decision are obtained from the natural utterances. To compute the voicing decision errors, the manual annotated of voicing decisions are compared with the F_0 model-generated values. The voicing/unvoicing decision errors obtained from the synthesized speech using three pitch tracking methods are shown in Fig. 3.9. In the figure, M1, M2, and M3 indicate RAPT, STRAIGHT, and proposed methods, respectively. From the results, it can be observed that the proposed method has significantly less v⇒u and u⇒v errors compared with RAPT and STRAIGHT methods. Among two errors, v⇒u error is less than u⇒v error for all three pitch tracking methods. The speech utterances of male speakers have slightly higher v⇒u and u⇒v errors compared with female speakers. Among all speakers, the KSP speaker has the highest v⇒u and u⇒v errors for RAPT and STRAIGHT methods. The main reason of high v⇒u and u⇒v errors is that among all speakers, the KSP speaker has highest percentage of creaky regions (about 2.7%, details will be discussed in Chap. 6), and the voicing detection performance of RAPT and

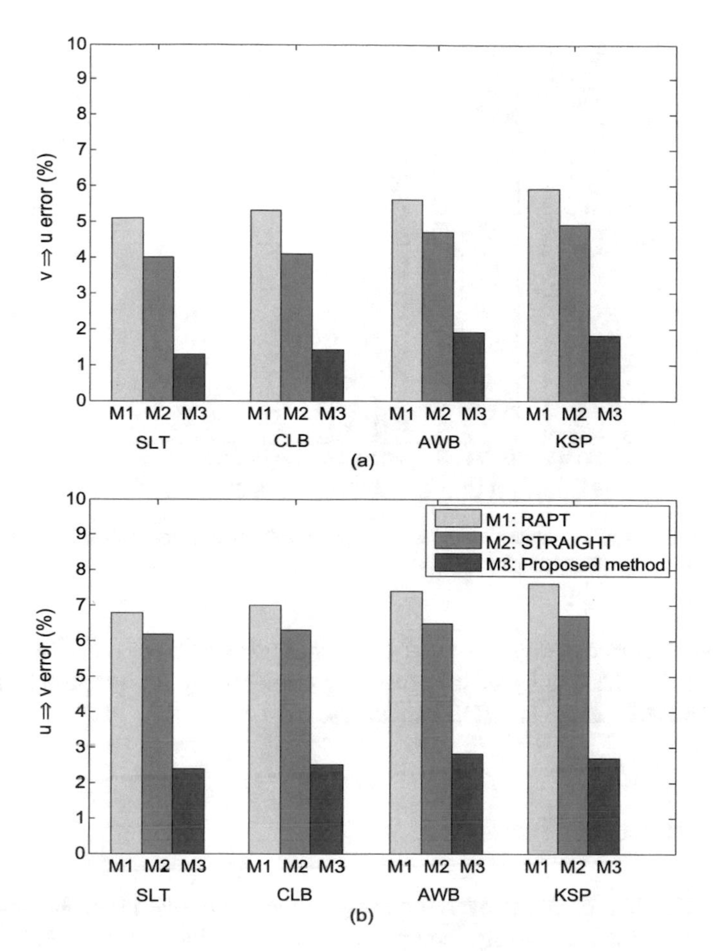

Fig. 3.9 (**a**) v⇒u and (**b**) u⇒v errors obtained for four speakers using three pitch tracking methods

STRAIGHT methods tends to degrade in creaky regions. The results also confirm that the low VDE (as mentioned in Sect. 3.2.5) obtained during the evaluation of the proposed pitch estimation method ensures reduction in v⇒u and u⇒v errors.

To further understand the nature of voicing/unvoicing decision errors, only phones having voiced/unvoiced decision errors are analyzed. Usually, the number of states having v⇒u and u⇒v errors in a phone may be one, two, or three. From the group of phones having v⇒u and u⇒v errors, the percentage of phones having the error in one state, two states, and three states are computed and are shown in Fig. 3.10. From the results, it can be observed that the percentage of phones having the error in two or three states is more in RAPT and STRAIGHT methods, compared to proposed method. This indicates that the voicing/unvoicing decision errors due to the proposed method will mostly occur at only one state of a phone. If the voicing/unvoicing decision errors occur in more than one state of a phone, then

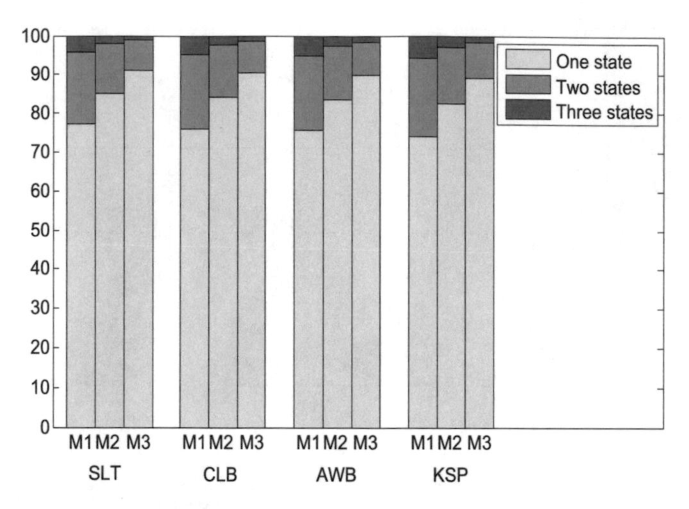

Fig. 3.10 Percentage of phones having error in one state, two states, and three states for four speakers using three pitch tracking methods

distortions are prominently perceived. From these observations, it can be objectively concluded that the quality of the speech synthesized by the proposed method is better than RAPT and STRAIGHT methods.

3.4.2 Subjective Evaluation

Subjective evaluation of the overall quality of the synthesized speech is performed using two measures, namely, comparative mean opinion scores (CMOS) and preference tests. In CMOS, the subjects were asked to listen to two versions of the synthesized speech; the first one is synthesized using the STRAIGHT or RAPT method, and the second one is synthesized using the proposed method. The subjects were asked to grade the overall preference between two sets of synthesized speech on a 7-point scale. The 7-point scale used to grade the preference between pairs of synthesized speech is shown in Table 3.3. A positive score indicates that the proposed method is preferred over other methods, and a negative score implies the opposite. In preference tests, the subjects were asked to give the preference between a pair of synthesized speech utterances. The subjects had the option either to prefer one of the synthesized speech utterances or to prefer both as equal. The synthesized speech files were played to the subjects in random sequence to avoid the bias toward any specific method. Listening tests were conducted with 20 research scholars in the age group of 23–35 years. The subjects have sufficient speech knowledge for proper assessment of the speech signals, as all of them have taken a full semester course on speech technology. Each of the subjects was given a pilot test about the perception of speech signals by playing the samples of synthesized speech files.

Table 3.3 Grades used in the CMOS test

Score	Subjective perception
3	Much better
2	Better
1	Slightly better
0	About the same
−1	Slightly worse
−2	Worse
−3	Much worse

Once they were comfortable with judging, they were allowed to take the tests. The tests were conducted in the laboratory environment by playing the speech signals through headphones. Before listening tests, the energies of the speech signals are normalized to the same level. The speech database, evaluation metrics, number of subjects, and the procedure followed for evaluation are the same for all subjective evaluation tests carried out in this work.

The CMOS scores with 95% confidence intervals and the preference scores are provided in Figs. 3.11 and 3.12, respectively. In figures, Set1 and Set2 indicate the comparison of the proposed method with the RAPT-F0-based HTS and the STRAIGHT-F0-based HTS, respectively. On comparison of the proposed method with the RAPT-F0-based HTS, it can be observed from Figs. 3.11 and 3.12 that the CMOS scores are varying between 1.5 and 2, and about 70–85% of the subjects preferred the proposed method. The above observation is true for both male and female speakers. Both CMOS and preference scores indicate that the proposed method is significantly better than RAPT-F0 based HTS. The subjects noticed that the monotonicity present in the speech synthesized by the RAPT-F0-based HTS is reduced in the speech synthesized by the proposed method. Subjects also identified that the speaker-specific characteristics are enhanced by the proposed method, compared to RAPT-F0-based HTS.

Regarding the comparison of the proposed method with the STRAIGHT-F0-based HTS, it can be noticed from Fig. 3.11 that the CMOS scores are varying between 0.6 and 1 for both male and female voices. The preference scores provided in Fig. 3.12 show that the subjects preferred the proposed method for about 40–45% and about 40% of the cases, the subjects suggested that both methods are equivalent. This shows that the speech synthesized by the proposed method is better than the STRAIGHT-F0-based HTS. On further observation of the synthesized speech, it can be noticed that, if the voicing/unvoicing decision errors (mostly v⇒u errors) are very less, or almost nil, then both the proposed and STRAIGHT methods are perceptually almost equivalent. The major difference between the proposed and STRAIGHT is the number of voicing/unvoicing decision errors generated in a sentence. The subjects could clearly identify the hoarseness or roughness at certain places in the speech synthesized by STRAIGHT-F0-based HTS. Synthesized speech samples of the proposed and two existing F_0 estimation methods are

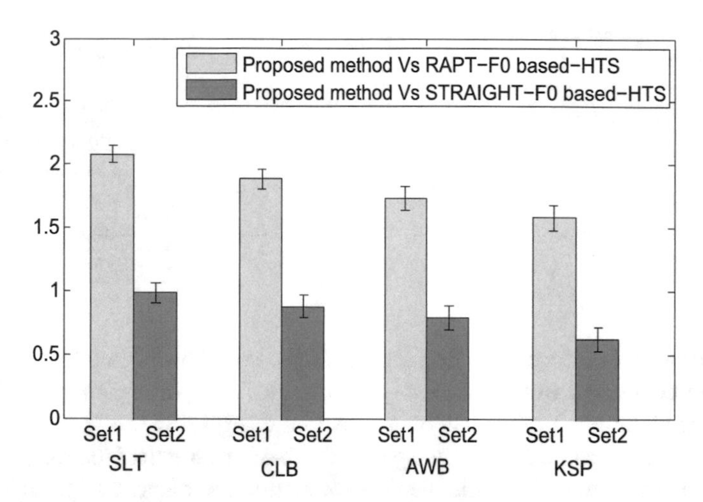

Fig. 3.11 CMOS scores with 95% confidence intervals obtained by comparing the proposed method with the RAPT-F0-based HTS and STRAIGHT-F0-based HTS

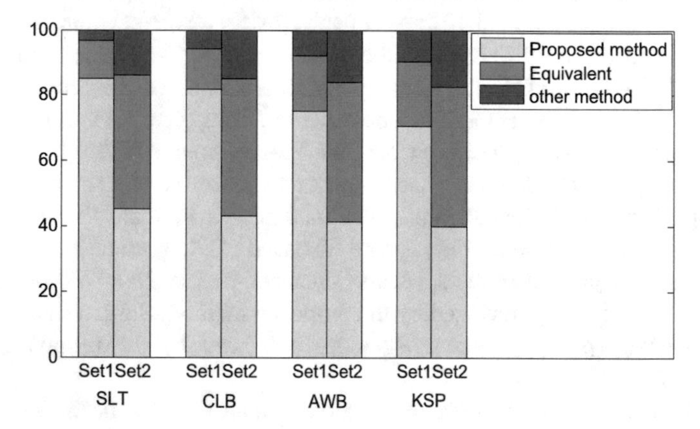

Fig. 3.12 Preference scores obtained by comparing the proposed method with the RAPT-F0-based HTS and STRAIGHT-F0-based HTS

made available online at http://www.sit.iitkgp.ernet.in/~ksrao/narendra-phd-demo/narendra_hsm.html.

3.5 Summary

This chapter proposed an efficient voicing detection and F_0 estimation method for HMM-based speech synthesis system. The voicing detection is performed based on the strength of instant significant excitation. The strength of excitation is efficiently computed from the ZFF method by choosing adaptive window length.

For the F_0 estimation, frame-wise zero-frequency filtering of speech is proposed. The performance of proposed method is compared with the existing pitch tracking methods. The performance evaluation results show that the voicing decision error and the gross pitch error of the proposed method are significantly low compared with other methods. The proposed voicing detection and F_0 extraction method was integrated into HMM-based speech synthesis. The quality of the synthesized speech from the proposed method is compared with two HMM-based speech systems developed using RAPT and STRAIGHT pitch estimation methods. The objective evaluation of voicing/unvoicing detection in HTS shows that the proposed method has significantly reduced voicing/unvoicing decision errors compared with RAPT and STRAIGHT methods. Subjective evaluation results showed a significant improvement in the preference for the proposed method over RAPT and STRAIGHT methods.

References

1. K. Tokuda, T. Yoshimura, T. Masuko, T. Kobayashi, T. Kitamura, Speech parameter generation algorithms for HMM-based speech synthesis, in *Proceedings of the International Conference on Acoustics, Speech, and Signal Processing (ICASSP)* (2000), pp. 1315–1318
2. K. Tokuda, T. Mausko, N. Miyazaki, T. Kobayashi, Multi-space probability distribution HMM. IEICE Trans. Inf. Syst. **E85-D**(3), 455–464 (2002)
3. J. Yamagishi, Z. Ling, S. King, Robustness of HMM-based speech synthesis, in *Proceedings of the Interspeech* (2008), pp. 581–584
4. D. Talkin, A robust algorithm for pitch tracking (RAPT), in *Speech Coding and Synthesis* (Elsevier Science, Amsterdam, 1995), pp. 495–518
5. P. Boersma, Accurate short-term analysis of fundamental frequency and the harmonics-to-noise ratio of a sampled sound. Inst. Phon. Sci. **17**, 97–110 (1993)
6. R. Goldberg, L. Riek, *A Practical Handbook of Speech Coders* (CRC Press, Boca Raton, 2000)
7. K.S.R. Murty, B. Yegnanarayana, Epoch extraction from speech signals. IEEE Trans. Audio Speech Lang. Process. **16**(8), 1602–1613 (2008)
8. P. Alku, T. Bakstrom, E. Vikman, Normalized amplitude quotient for parameterization of the glottal flow. J. Acoust. Soc. Am. **112**(2), 701–710 (2002)
9. K.S.R. Murty, B. Yegnanarayana, M.A. Joseph, Characterization of glottal activity from speech signals. IEEE Signal Process. Lett. **16**(6), 469–472 (2009)
10. B. Yegnanarayana, K.S.R. Murty, Event-based instantaneous fundamental frequency estimation from speech signals. IEEE Trans. Audio Speech Lang. Process. **17**(4), 614–624 (2009)
11. Y. Bayya, D.N. Gowda, Spectro-temporal analysis of speech signals using zero-time windowing and group delay function. Speech Commun. **55**(6), 782–795 (2013)
12. D.J. Hermes, Measurement of pitch by subharmonic summation. J. Acoust. Soc. Am. **83**(1), 257–264 (1988)
13. H. Kawahara, H. Katayose, A. de Cheveigne, R. Patterson, Fixed point analysis of frequency to instantaneous frequency mapping for accurate estimation of F0 and periodicity, in *Proceedings of the Eurospeech* (1999), pp. 2781–2784
14. T. Drugman, A. Alwan, Joint robust voicing detection and pitch estimation based on residual harmonics, in *Proceedings of the Interspeech* (2011), pp. 1973–1976
15. F. Plante, G.F. Meyer, W.A. Aubsworth, A pitch extraction reference database, in *Eurospeech* (1995), pp. 837–840

16. P. Bagshaw, S.M. Hiller, M.A. Jack, Enhanced pitch tracking and the processing of FQ contours for computer and intonation teaching, in *Eurospeech* (1993), pp. 1003–1006
17. HMM-based speech synthesis system (HTS). Available: http://hts.sp.nitech.ac.jp/
18. J.J. Odella, The use of context in large vocabulary speech recognition, Ph.D. dissertation, Cambridge University, 1995
19. K. Shinoda, T. Watanabe, MDL-based context-dependent subword modeling for speech recognition. J. Acoust. Soc. Jpn. (E) **21**(2), 79–86 (2000)
20. T. Toda, K. Tokuda, A speech parameter generation algorithm considering global variance for HMM-based speech synthesis. IEICE Trans. Inf. Syst. **90**(5), 816–824 (2007)
21. CMU ARCTIC speech synthesis databases. Available: http://festvox.org/cmu_arctic/
22. H. Zen, T. Toda, M. Nakamura, K. Tokuda, Details of Nitech HMM-based speech synthesis system for the Blizzard Challenge 2005. IEICE Trans. Inf. Syst. **E90-D**(1), 325–333 (2007)
23. H. Zen, T. Toda, K. Tokuda, The Nitech-NAIST HMM-based speech synthesis system for the Blizzard Challenge 2006. IEICE Trans. Inf. Syst. **E91-D**(6), 1764–1773 (2008)
24. K. Oura, H. Zen, Y. Nankaku, A. Lee, K. Tokuda, A tied covariance technique for HMM-based speech synthesis. IEICE Trans. Inf. Syst. **E93-D**(3), 595–601 (2010)
25. Q. Zhang, F. Soong, Y. Qian, Z. Yan, J. Pan, Y. Yan, Improved modeling for F0 generation and V/U decision in HMM-based TTS, in *Proceedings of the International Conference on Acoustics Speech and Signal Processing (ICASSP)* (2010), pp. 4606–4609

Chapter 4
Parametric Approach of Modeling the Source Signal

4.1 Parametric Source Modeling Method Based on Principal Component Analysis

The proposed source modeling method models the pitch-synchronous residual frames extracted from the excitation signal. Initially, energy is extracted from every frame of the excitation signal. Then, the pitch-synchronous analysis is performed on the excitation signal leading to a set of residual frames that are synchronous with the GCI and whose length is set to two pitch periods (explained in Sect. 4.1.1). Using the generated pitch-synchronous residual frames, principal component analysis is performed. Based on the analysis, every residual frame is parameterized in terms of PCA coefficients (explained in Sect. 4.1.2). Frame-wise energy and PCA coefficients are considered as the excitation parameters, and these parameters are modeled under HMM framework.

At the time of synthesis, the excitation parameters are generated from the HMMs. From the generated PCA coefficients, the residual frames are reconstructed (described in Sect. 4.1.3). The reconstructed residual frames are overlap-added to generate the excitation signal. The steps involved in modeling of excitation signal based on PCA are explained in the following section.

4.1.1 Generation of Pitch-Synchronous Residual Frames

The sequence of steps followed in the generation of pitch-synchronous residual frames is shown in Fig. 4.1. From every speech utterance present in the database, mel-generalized cepstrum (MGC) coefficients are extracted. MGC coefficients capture the spectral envelope of speech [1]. As suggested in [2], 34th order MGC coefficients are extracted with the parameter values $\alpha = 0.42$ ($Fs = 16\,\text{kHz}$)

K. S. Rao, N. P. Narendra, *Source Modeling Techniques for Quality Enhancement in Statistical Parametric Speech Synthesis*, SpringerBriefs in Speech Technology,
https://doi.org/10.1007/978-3-030-02759-9_4

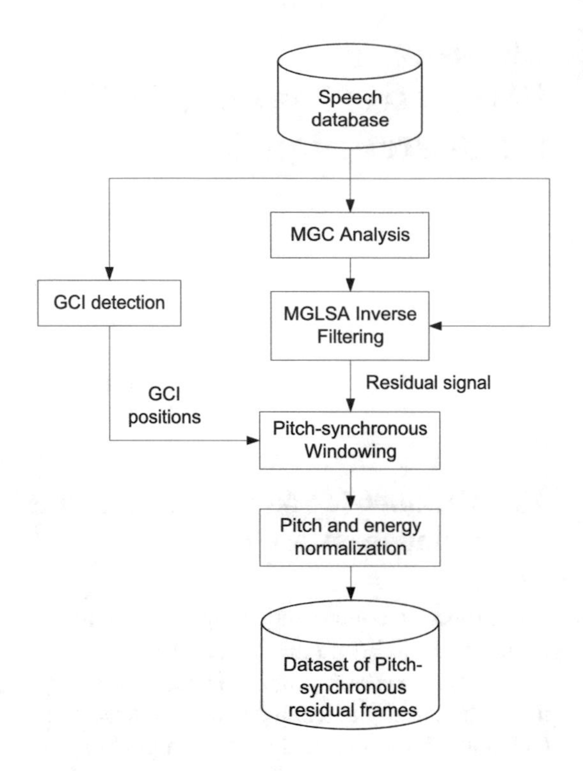

Fig. 4.1 Flow diagram for obtaining the pitch-synchronous residual frames

and $\gamma = -1/3$. The excitation or residual signal is obtained by inverse filtering using mel-generalized log-spectral approximation (MGLSA) digital filter. GCIs are estimated from the speech using zero frequency filtering method [3]. The main reason for choosing ZFF method is that it has good identification rate and accuracy. The procedure for extracting the locations of GCIs from the speech signal using ZFF method is provided in Sect. 3.2.1. The GCIs are particular events in speech production which provide significant excitation to the vocal-tract system. Using the knowledge of GCI positions, the boundaries of pitch cycles are marked on the excitation signal. Using GCI positions as anchor points, two-pitch period long residual signals are extracted, and they are Hanning windowed. During the extraction of the residual signal, it is ensured that the GCI is at the center of the residual frame. The extracted residual signals are normalized both in pitch period and energy. In this work, the normalized pitch value is chosen as small as possible, such that during synthesis most of the times the pitch-normalized residual frames are downsampled according to the target pitch values. During synthesis, upsampling the pitch-normalized residual frame may result in the appearance of energy holes at high frequency [4]. Here, upsampling is performed to increase the number of samples (in the case of decreasing the pitch) in the residual frame, and downsampling is carried out to decrease the number of samples (in the case of increasing the pitch) in the residual frame. On analyzing the speaker's pitch histogram ($P(F_0)$), the chosen normalized pitch value F_0^* typically satisfies the following condition:

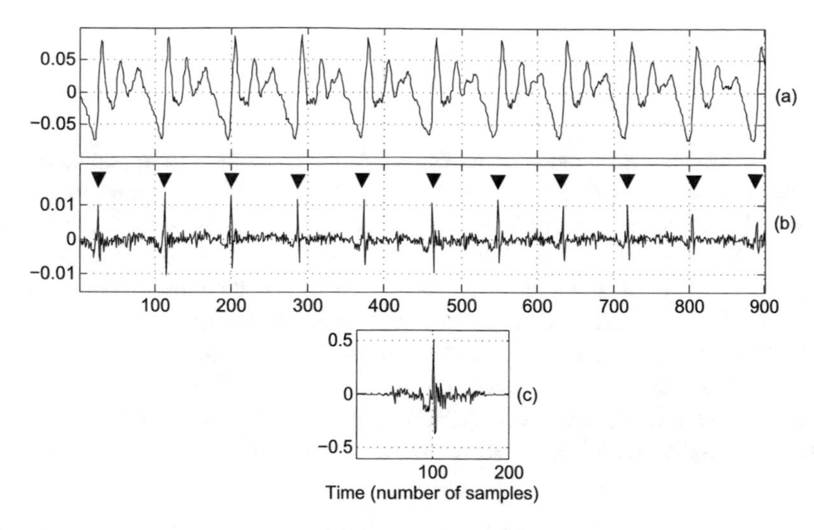

Fig. 4.2 (**a**) Speech signal, (**b**) residual signal, and (**c**) pitch-synchronous residual frame. The locations of GCIs are shown by downward arrows

$$\int_{F_0^*}^{\infty} P(F_0)\,\mathrm{d}F_0 \geq 0.9 \tag{4.1}$$

such that less than 10% frames will be upsampled during synthesis time. The pitch is estimated using the voicing detection and F_0 estimation method proposed in Chap. 3. The energy normalization is achieved by fixing the total energy of the residual frame to 1. These operations ensure proper alignment of the residual frames and avoid the estimation errors due to mismatch of GCI locations. The resulting residual frames are comparable so that they can be analyzed under a common framework. In this work, GCI-centered two-pitch period long and Hanning windowed residual signals are viewed as the pitch-synchronous residual frames. Figure 4.2a–c shows the segment of the speech signal, its corresponding residual signal, and an example of extracted pitch-synchronous residual frame, respectively. The locations of GCIs are shown by downward arrows in Fig. 4.2b.

4.1.2 Parameterization of Residual Frame Using PCA

From the pitch-synchronous residual frames extracted from the excitation signal, principal component analysis is performed. The principal component analysis is widely used in various approaches such as dimensionality reduction, lossy data compression, and feature extraction [5, 6]. For PCA, a covariance matrix is constructed from the statistics of the given data vectors [7]. Here, the data vectors correspond to the waveforms of residual frames. Using PCA, the residual frame (**x**) can be reconstructed as follows:

$$\tilde{\mathbf{x}} = \sum_{n=1}^{N} \alpha_n \mathbf{u}_n + \bar{\mathbf{x}} \tag{4.2}$$

where $\tilde{\mathbf{x}}$ is the reconstructed residual frame, N is the number of eigenvectors, and $\bar{\mathbf{x}}$ is the sample mean of $\mathbf{x}$. $\mathbf{u}_n$ is the eigenvector of the covariance matrix $\sum_{\mathbf{x}} = E\{\mathbf{x}\mathbf{x}^T\}$, and α_n is the coefficient associated with $\mathbf{u}_n$. It is assumed that the eigenvectors are ordered according to the eigenvalues $\lambda_1 > \lambda_2 > \ldots > \lambda_N$. λ_n is known to represent the data dispersion along $\mathbf{u}_n$. For compact representation, only first $N' < N$ eigenvectors or principal components are used which results in N' PCA coefficients. The principal components represent the directions of the largest variance in the signal space. By using the first N' principal components, the cumulative relative dispersion (CRD) ratio is defined as the ratio of variance represented in the first N' eigenvectors to the total variance.

$$CRD(N') = \frac{\sum_{i=1}^{N'} \lambda_i}{\sum_{i=1}^{N} \lambda_i} \tag{4.3}$$

For PCA, 10,000 pitch-synchronous residual frames extracted from SLT speaker of CMU Arctic database are considered [8]. Figure 4.3 displays a typical evolution of cumulative relative dispersion for different numbers of eigenvectors. From the figure, it can be observed that to represent entire residual frame accurately, 120 eigenvectors are required. Using 120-dimensional PCA coefficients to represent every residual frame can lead to the problem of "curse of dimensionality." From the figure, it can be observed that with the increase in the number of eigenvectors, the CRD value is increasing exponentially up to 15 eigenvectors. From 15 to 30 eigenvectors, the CRD value is increasing linearly with the number of eigenvectors. From 30 to 120 eigenvectors, the CRD value is increasing slowly (almost logarithmically) with the number of eigenvectors. Also, it can be observed that the contribution of lower eigenvectors is significantly more compared to higher eigenvectors. Instead

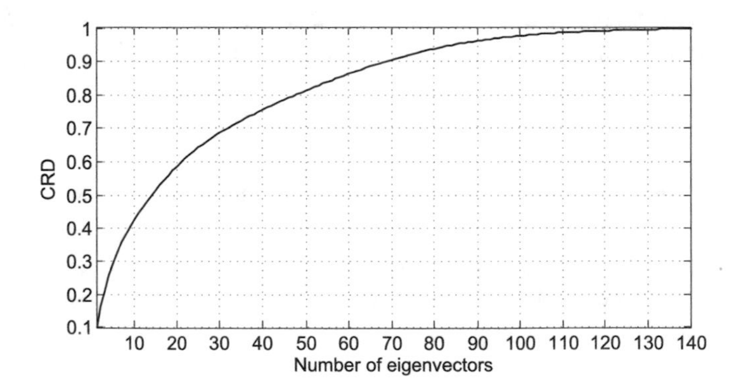

Fig. 4.3 Evolution of cumulative relative dispersion (CRD) as a function of number of eigenvectors for SLT speaker. Total number of eigenvectors = 200

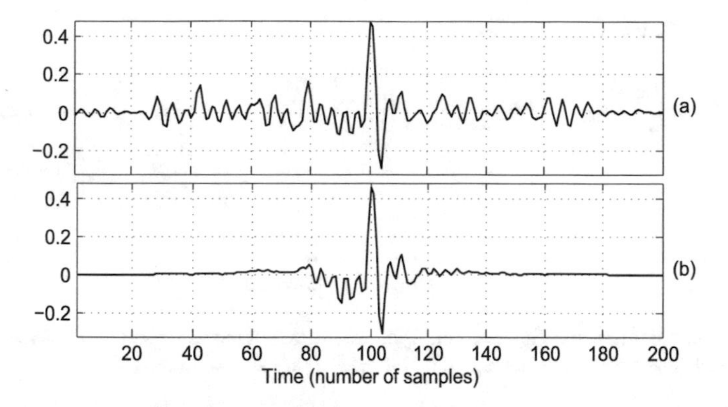

Fig. 4.4 (**a**) Original residual frame and (**b**) reconstructed residual frame using 30 PCA coefficients and eigenvectors

of representing the entire residual frame, major portion of the residual frame can be represented by using the first 30 eigenvectors. From Fig. 4.3, it can be observed that the first 30 eigenvectors represent up to 69% of the variance. Every residual frame is projected on the first 30 eigenvectors to obtain 30 PCA coefficients. Figure 4.4a, b shows the original residual frame and the reconstructed residual frame using 30 PCA coefficients and eigenvectors, respectively.

In HTS, both MGC coefficients (34th order with $\alpha = 0.42$, $Fs = 16\,\text{kHz}$, and $\gamma = -1/3$) and F_0 values are extracted at a frame size of 25 ms with a frame shift of 5 ms. The excitation parameters which include PCA coefficients are extracted from every pitch-synchronous residual frames. As it is convenient to model all parameters extracted at a constant frame size and frame rate in a unified framework, the PCA coefficients extracted from the pitch-synchronous residual frames present in every 25 ms frame are averaged and assigned as the parameters of that frame. In addition, energy is extracted from every frame of the excitation signal. In the case of unvoiced speech, only frame-wise energy is extracted from the excitation signal, and PCA coefficients are set to zero. Except F_0, all other parameters are modeled by CD-HMMs. F_0 patterns are modeled by MSD-HMMs. Other specifications of training HMMs are same as mentioned in Chap. 3. During synthesis, for the given input text, MGC coefficients and excitation parameters are generated from the HMMs. From the generated excitation parameters, the excitation signal is constructed which is used to synthesize the speech.

4.1.3 Speech Synthesis Using the Proposed PCA-Based Parametric Source Model

The block diagram showing different synthesis stages is shown in Fig. 4.5. In the figure, the parameters generated from the HMMs are shown in italics. The excitation signal is generated separately for voiced and unvoiced frames. For voiced frame,

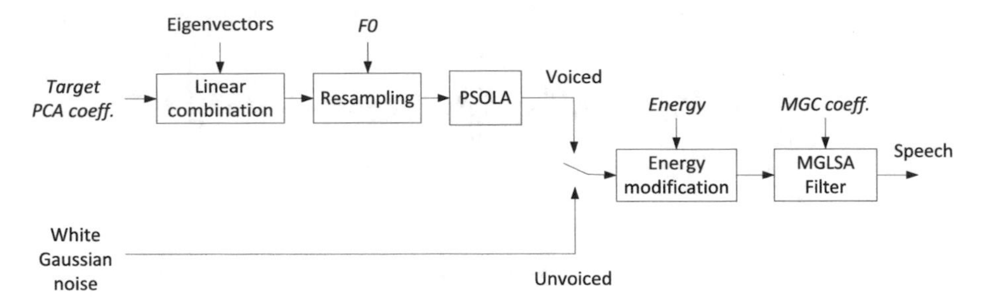

Fig. 4.5 Block diagram showing different synthesis stages in the proposed PCA-based parametric source model. The parameters generated by the HMMs are shown in italics

the residual frame is reconstructed from the linear combination of eigenvectors and target PCA coefficients. The residual frame is resampled to twice the target pitch period. The residual frames are pitch-synchronously overlap-added to generate the excitation signal. In all source models proposed in this book, the time-domain pitch-synchronous overlap-add (TD-PSOLA) method is used for constructing the excitation signal. In TD-PSOLA, 50% of each of residual frames are overlapped with the adjacent frames. During overlap-add, it is ensured that the maxima of residual frames are matched with the glottal closure instants which are accurately estimated by using ZFF method. These steps ensured the phase continuity in the excitation signal [9]. The energy of the resulting excitation signal is modified according to the energy measure generated from the HMM. For unvoiced speech, white noise whose energy is modified according to the generated energy measure is used as excitation signal. The resulting excitation signal is given as input to the MGLSA filter, controlled by MGC coefficients to generate the speech.

4.1.4 Evaluation

The proposed method is evaluated using four speakers (SLT, CLB, AWB, and KSP) from CMU Arctic speech database [8]. Subjective evaluation is performed using two measures, namely, CMOS and preference tests. The details about the speech database, evaluation metrics, number of subjects, and the procedure followed for evaluation are provided in Chap. 3. The quality of synthesized speech from the proposed method is compared with Pulse-HTS. In Pulse-HTS, a sequence of pulses positioned according to the generated pitch is used as the excitation signal. Pulse-HTS is known for its simple excitation scheme, and it is mostly used as a reference for testing the proposed methods.

Figure 4.6a, b provides the CMOS scores with 95% confidence intervals and the preference scores obtained by comparing the proposed method with Pulse-HTS. The CMOS scores of both female and male speakers are greater than zero which indicates that the proposed method is better than the existing methods. The CMOS

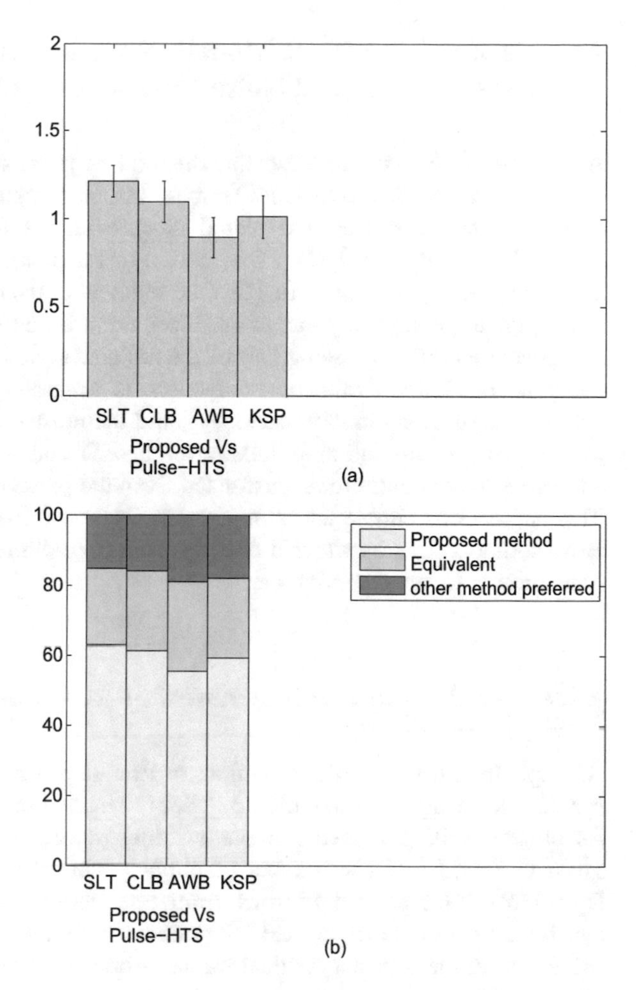

Fig. 4.6 (**a**) CMOS scores with 95% confidence intervals and (**b**) preference scores obtained by comparing the proposed method with Pulse-HTS

scores of the female speakers are relatively higher compared to the male speakers. For both female and male speakers, the subjects preferred the proposed method compared to Pulse-HTS.

From the subjective evaluation results, it is observed that the proposed parametric approach based on PCA coefficients is better than the Pulse-HTS. The quality of the synthesized speech is still far from natural quality, and there exists some amount of buzziness in the synthesized speech. In the proposed parametric method, every residual frame is parameterized using 30 PCA coefficients which represent only 69% of the residual frame. To accurately represent every residual frame, the second parametric source modeling method is proposed which is based on the analysis of characteristics of the residual frames around GCI.

4.2 Parametric Source Modeling Method Based on the Deterministic and Noise Components of Residual Frames

A new parametric source modeling method is proposed based on the analysis of characteristics of the residual frames. Initially, a study on the characteristics of the residual signal around glottal closure instant (GCI) is performed using principal component analysis (PCA). Based on the present study, and from previous literature [10, 11], it is concluded that the segment of the residual signal around GCI which carries perceptually important information is considered as the deterministic component and the remaining part of the residual signal is considered as the noise component. The deterministic component is accurately represented using PCA coefficients (with about 95% accuracy), and the noise component is parameterized in terms of spectral and amplitude envelopes. During synthesis, the deterministic and noise components are reconstructed from the parameters generated by HMMs. This approach is simple and computationally less intensive, as deterministic and noise components are extracted directly from time-domain representation, without transforming to any other domain.

4.2.1 Analysis of Characteristics of Residual Frames

As explained in the previous section, at first step, the excitation signal is pitch-synchronously decomposed into a number of residual frames. The sequence of steps for obtaining the pitch-synchronous residual frames from the excitation signal is given in Sect. 4.1.1. The number of pitch-synchronous residual frames extracted from the excitation signal varies from one phone to another. Adjacent pitch-synchronous residual frames exhibit strong correlation [12]. On close observation, the shapes of the adjacent residual frames around GCI are very much similar. Most of the existing approaches parameterize the entire residual frame by considering either the time-domain or frequency-domain representation of the signal [13–15]. They do not parameterize the residual frames based on its perceptual significance. In [10, 11], it is observed that in the entire residual frame, the region around GCI carries important information related to the perceptual characteristics of the voiced speech. Motivated by this observation, to further analyze the characteristics of the residual signal around GCI, principal component analysis is performed on the pitch-synchronous residual frames. For analysis, 10,000 residual frames extracted from SLT speaker of CMU Arctic database [8] are considered. For compact representation, each of the residual frames of dimension N can be projected on $N' < N$ eigenvectors which results in N' PCA coefficients. For a single residual frame, with the different numbers of eigenvectors ($N' = 5$, 10, 15, 20, and 25), PCA coefficients are computed. Using the different numbers of eigenvectors and PCA coefficients, the residual frames are reconstructed, and their variations are analyzed. Figure 4.7 shows the original residual frame and the residual frames reconstructed using first 5, 10, 15, 20, and 25 eigenvectors. From the figure, it can be observed that by considering lower-order eigenvectors (5 and 10), only the region around GCI (middle portion of the residual frame) is reconstructed. Finer details present at other regions are captured, as the order of eigenvectors is increased. On observation of the evolution of cumulative relative dispersion for different numbers of eigenvectors shown in Fig. 4.3, it can be seen that about 59% of the variance is represented by the first 20 eigenvectors which mainly corresponds to the region around GCI of the residual frame. To represent the remaining part of the residual frame, 100 higher-order eigenvectors are required. The region around GCI represents most of the variance and hence can be regarded as the dominant part of the residual frame. Drugman et al. [16] and Drugman and Dutoit [14] have shown that the segment of the residual signal around GCI is closely related to LF model [17].

Based on the above observation, the residual signal can be divided into two parts. The first part is the small segment of the residual signal around GCI, and the second part is the remaining segment of the residual signal. The segment of the residual signal around GCI is considered to have equal length on either side of GCI. To ensure smooth continuity at the joining points, the small segment of the residual signal around GCI is Hanning windowed. The Hanning windowed segment was subtracted from the residual frame to obtain the second part. The first part can be predicted from a small number of eigenvectors (about 20), and hence it can be considered as the deterministic component of the residual frame. The second part (i.e., other than deterministic component) requires a large number of eigenvectors (about 100) for accurate estimation, and hence it can be considered as the noise component of the residual frame. Figure 4.8 provides the deterministic and noise components extracted from the residual frame shown in Fig. 4.7a.

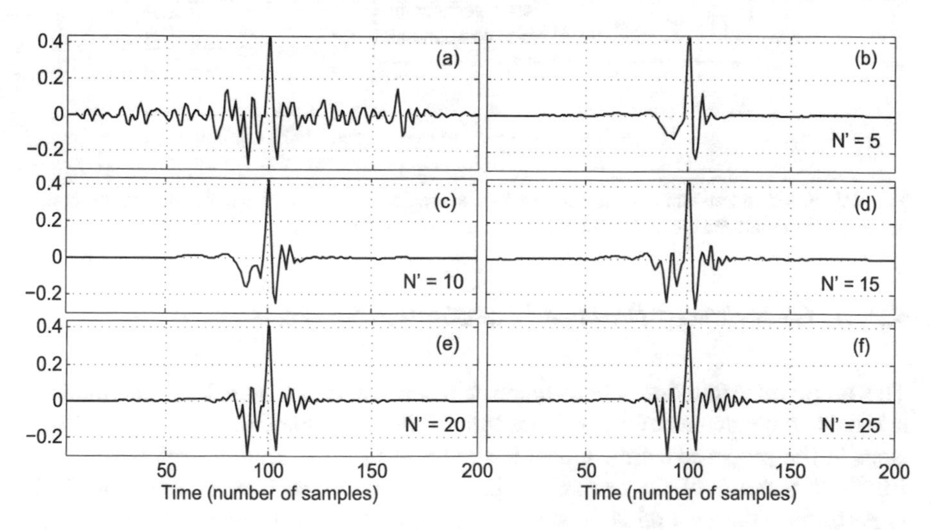

Fig. 4.7 (**a**) Original residual frame. Residual frame reconstructed using (**b**) 5, (**c**) 10, (**d**) 15, (**e**) 20, and (**f**) 25 eigenvectors

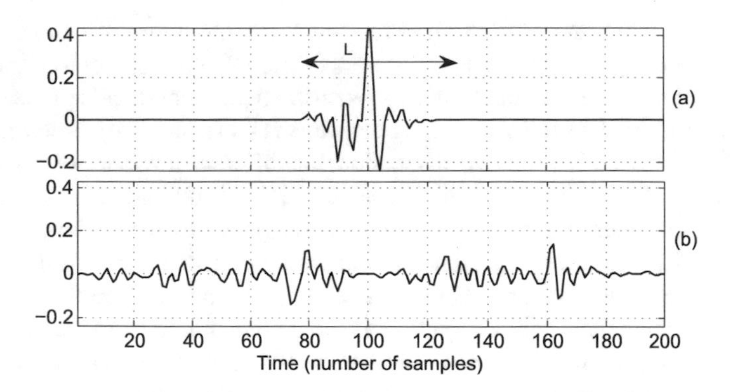

Fig. 4.8 (**a**) Deterministic and (**b**) noise components extracted from the residual frame given in Fig. 4.7a

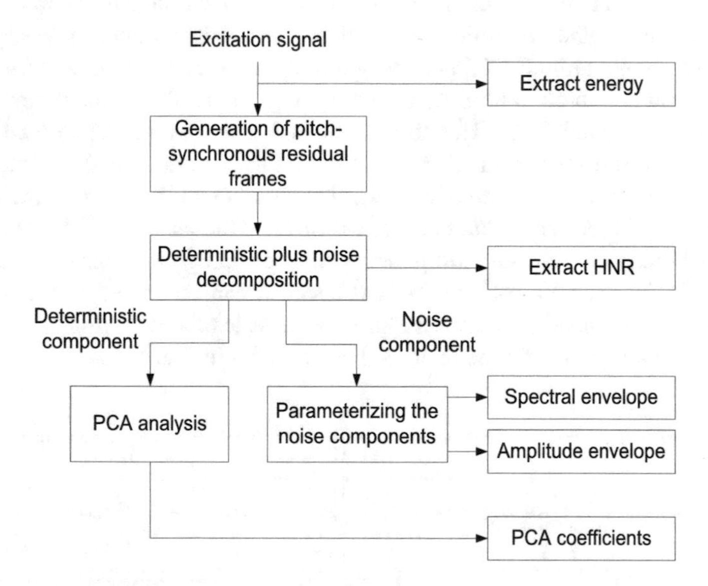

Fig. 4.9 Flowchart indicating the sequence of steps in the proposed parametric source model based on the deterministic and noise components of residual frames

4.2.2 *Overview of Proposed Parametric Source Model*

The proposed excitation model represents the excitation signal as deterministic and noise components of the residual signal. The flow diagram indicating different steps in the proposed source model is shown in Fig. 4.9. First, energy is extracted from every frame of the excitation signal. Then, the pitch-synchronous analysis is performed on the excitation signal leading to a set of residual frames that are synchronous with the GCI and whose length is set to two pitch periods (described in Sect. 4.1.1). From the pitch-synchronous residual frames, deterministic and noise components are computed using the proposed approach. The deterministic component is accurately represented using 20 PCA coefficients (explained in Sect. 4.2.3), and the noise component is parameterized in terms of spectral and amplitude envelopes (explained in Sect. 4.2.4). Harmonic to noise ratio (HNR) is computed as the ratio of the energy of deterministic and noise components. Energy, PCA coefficients, HNR, and spectral and amplitude envelopes are considered as excitation parameters. At the time of synthesis, the deterministic component waveform is reconstructed from the generated PCA coefficients, and the noise component is obtained by imposing the target spectral and amplitude envelopes on the white Gaussian noise. The deterministic and the noise components are pitch-synchronously overlap-added to generate the excitation signal (described in Sect. 4.2.5).

4.2.3 Parameterization of Deterministic Component

Before parameterizing the deterministic component, the length of deterministic component (L), i.e., the length of the segment of the residual signal around GCI as shown in Fig. 4.8a, should be fixed. The length should be appropriately chosen such that the deterministic component can be accurately represented with M number of eigenvectors. First, by varying the length L from 2 to twice the normalized pitch period (in the number of samples) in steps of 2 samples, the deterministic components are extracted from the residual frames. Here, 10,000 residual frames from SLT speaker are considered. By considering the deterministic components of every length L, PCA is performed. For every L, the CRD value is computed for M number of eigenvectors. The largest possible L which results in CRD value $\geq$95% is considered as the appropriate length of the deterministic component. Choosing L with CRD value $\geq$95% ensures accurate representation of the deterministic component.

Before finding the appropriate length of the deterministic component, the number of eigenvectors M used for representing the deterministic component should be fixed. By varying M from 1 to 200, the length of the deterministic component is computed which results in the CRD value $\geq$95%. Increasing the value of M results in the subsequent increase in the value of L and vice versa. If M is chosen very small, the length L will also be very small. This may not exactly capture the region around GCI and results in reduced quality of speech. If M is chosen very large, then the complexity of model increases, and more data is required to capture the actual distribution. For $M = 20$, the length of the deterministic component is observed to be optimal (about one-third the length of the residual frame). Hence in this study, M is fixed to 20.

With $M = 20$, CRD values computed for different lengths of deterministic components are shown in Fig. 4.10. The CRD value is close to 100% for smaller lengths of deterministic components. From the figure, it can be observed that the largest possible L with CRD value $\geq$95% is 56. With $L = 56$, the deterministic components are extracted from the residual frames of SLT speaker, and PCA is performed. Each deterministic component is compactly represented by using 20

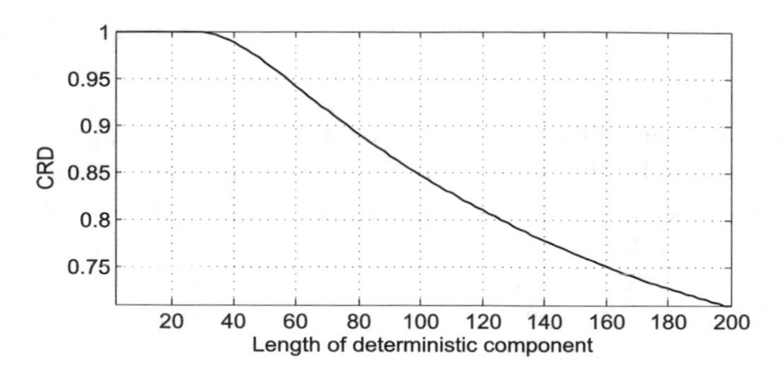

Fig. 4.10 Cumulative relative dispersion (CRD) values computed for different lengths of deterministic component (L) for SLT speaker

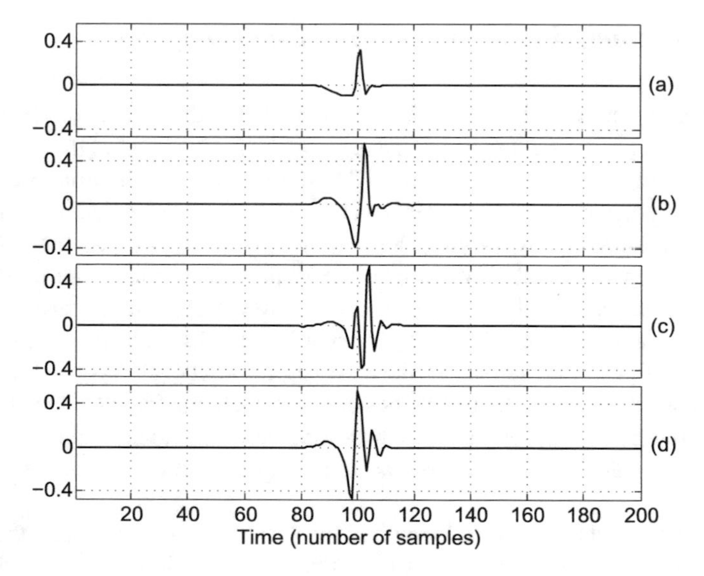

Fig. 4.11 (**a**) Mean vector, (**b**) first, (**c**) second, and (**d**) third eigenvector of the deterministic component

PCA coefficients. The deterministic component waveform mean vector and the first three eigenvectors are shown in Fig. 4.11. The mean vector captures the average shape of the deterministic component waveform, and other components model the rising and decaying patterns just before and after GCI.

4.2.4 Parameterization of Noise Component

The noise component is parameterized in terms of its spectral and amplitude envelopes. The spectral envelope of the noise component is estimated by using all-pole filtering or linear prediction (LP) coefficients. A typical criterion to select the order of the LP analysis is to use one complex pole per kHz of the total bandwidth (equal to half the sample rate) plus two to four additional poles [18]. For the sampling frequency of 16 kHz, 10–14 poles are typically used for LP analysis. Hence in this work, the order of LPC is chosen to be 10. The LPC coefficients are converted to LSF coefficients. The LSFs have better quantization properties and result in low spectral distortion than the conventional LPC coefficients [19, 20]. In literature, the LSFs have been proven to be best suitable in statistical parametric speech synthesis system [21].

The amplitude envelope ($a(n)$) is obtained by filtering the absolute value of noise component ($u(n)$) with a moving average filter of order $2N + 1$. N is chosen to be 8. The amplitude envelope is given by

$$a(n) = \frac{1}{(2N+1)} \sum_{k=-N}^{N} |u(n-k)|. \tag{4.4}$$

Normalization of the envelope is performed by setting the maximum value to 1. This method of amplitude envelope estimation was previously performed by Pantazis et al. [22]. Due to smoothening by the moving average filter, the amplitude envelope shows slow variation. The overall shape of the amplitude envelope is represented by a small number of samples. In our case, the amplitude envelope is represented by downsampling it into 15 samples. Increasing the sampling number increases the accuracy of representation of the amplitude envelope. The amplitude envelope is downsampled to a different number of samples, namely, 6, 9, 12, 15, 18, 21, and 24, and again upsampled back to the original normalized pitch value. The distance between the upsampled ($x_{ups}(n)$) and the original ($x_{org}(n)$) amplitude envelope is computed using relative time squared error ($RTSE$) measure. $RTSE$ is given by (where m is the number of samples of the normalized pitch)

$$RTSE = \frac{\sum_{i=1}^{m} (x_{ups}(i) - x_{org}(i))^2}{\sum_{i=1}^{m} x_{org}(i)^2} \tag{4.5}$$

Table 4.1 provides the RTSE values between the upsampled and original amplitude envelopes computed for a different number of samples. To calculate the RTSE values, 1000 instances of amplitude envelopes of noise signal are considered from AWB speaker of CMU Arctic database [8]. From the table, it can be observed that in above sample number of 15, RTSE is not changing significantly. Hence, 15 samples obtained after downsampling are used as the parameters representing the amplitude envelope of the noise component.

Similar to the previous method, the excitation parameters, namely, PCA coefficients, spectral and amplitude envelopes of the noise component, and HNR, computed from the pitch-synchronous residual frames present in every 25 ms frame are averaged and assigned as the parameters of that frame. In the case of unvoiced speech, except energy of excitation signal, all other excitation parameters are set

Table 4.1 Relative time squared error between the upsampled and original amplitude envelopes

Number of samples	RTSE
6	0.0959
9	0.0655
12	0.0385
15	0.0145
18	0.0132
21	0.0118

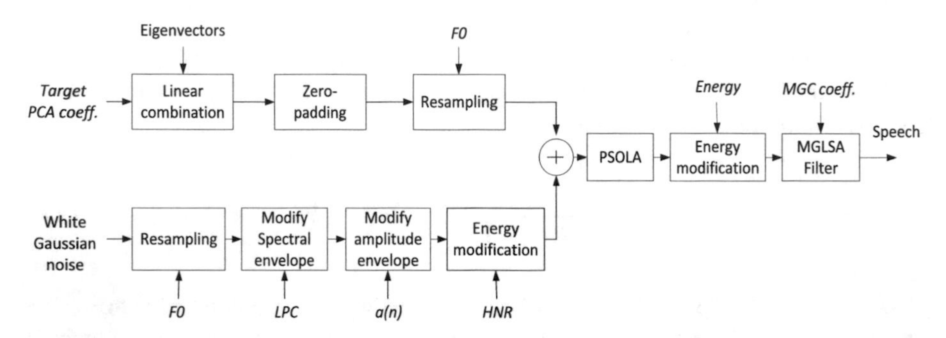

Fig. 4.12 Block diagram showing different synthesis stages in the proposed deterministic and noise component-based parametric source model. The parameters generated by the HMMs are shown in italics

to zero. Except F_0, all parameters including MGC coefficients (34th order with $\alpha = 0.42$, $Fs = 16\,\text{kHz}$, and $\gamma = -1/3$) are modeled by CD-HMMs. F_0 patterns are modeled by MSD-HMMs.

4.2.5 Speech Synthesis Using the Proposed Deterministic and Noise Component-Based Parametric Source Model

During synthesis, MGC coefficients, F0 including voicing decision and excitation parameters, are generated from HMMs by using constrained maximum likelihood algorithm [23]. The block diagram showing different synthesis stages is shown in Fig. 4.12. In the figure, the parameters generated by the HMMs are shown in italics. The excitation signal is generated separately for voiced and unvoiced frames. For the voiced frame, the deterministic component of the residual frame is obtained from the linear combination of eigenvectors and target PCA coefficients. The deterministic component is zero padded on either side such that its length is twice the normalized pitch period. The zero padded deterministic component is resampled to twice the target pitch period. The noise component of the residual frame is constructed by using white Gaussian noise. First, the white Gaussian noise is resampled to twice the target pitch period. Then, the target spectral envelope generated by the HMM is imposed on the resampled white Gaussian noise. The target spectral envelope is the all-pole model of noise represented by LSF coefficients. The LSFs are converted to LPCs (a_ks). An IIR filter is constructed which filters the white noise signal to obtain the desired target spectrum. The transfer function of IIR filter ($H(z)$) is given by

$$H(z) = \frac{1}{(1 - G(z))} \tag{4.6}$$

where $G(z) = \sum_{k=1}^{p} a_k z^{-k}$ is the FIR filter obtained from the LPCs of target spectral envelope. The target amplitude envelope ($a(n)$) generated by the HMM is imposed on the IIR filtered noise signal. The target amplitude envelope which is represented by 15 samples is upsampled to the required target pitch period. The amplitude envelope of the IIR filtered noise signal is also computed. The target envelope is imposed on the IIR filtered noise signal by compensating the difference between two envelopes. The energy of the spectrum and amplitude envelope modified noise signal is changed according to the generated HNR. Both deterministic and noise components are superimposed and then overlap-added to generate the excitation signal. The energy of the excitation signal is modified according to the energy measure generated by the HMM. For the unvoiced speech, the white noise whose energy is modified according to the generated energy measure is used as the excitation signal. The resulting excitation signal is given as input to the MGLSA filter, controlled by MGC coefficients to generate the speech.

4.2.6 Evaluation

The proposed method is evaluated using four speakers (SLT, CLB, AWB, and KSP) from CMU Arctic speech database [8]. Subjective evaluation is performed using two measures, namely, CMOS and preference tests. The details about the speech database, evaluation metrics, number of subjects, and the procedure followed for evaluation are provided in Chap. 3. The quality of the synthesized speech from the proposed method is compared with the following three existing excitation modeling methods.

1. **Pulse-HTS:** Pulse-HTS is known for its simple excitation scheme, and it is mostly used as a reference for testing the proposed methods. In Pulse-HTS, MGC coefficients are used to model the spectrum of speech, and pitch is used as the excitation feature. Figure 4.13 shows the block diagram indicating the sequence of steps in the traditional pulse/noise excitation. For voiced speech, a sequence of pulses positioned according to the generated pitch is used as the excitation signal. For unvoiced speech, white noise is used as the excitation signal. The speech is synthesized from the excitation signal and MGC coefficients using the MGLSA filter. The synthesized speech is smooth and contains typical buzziness.
2. **STRAIGHT-HTS [24]:** STRAIGHT-HTS is one of the most widely used methods for high-quality speech synthesis and uses mixed excitation parametric

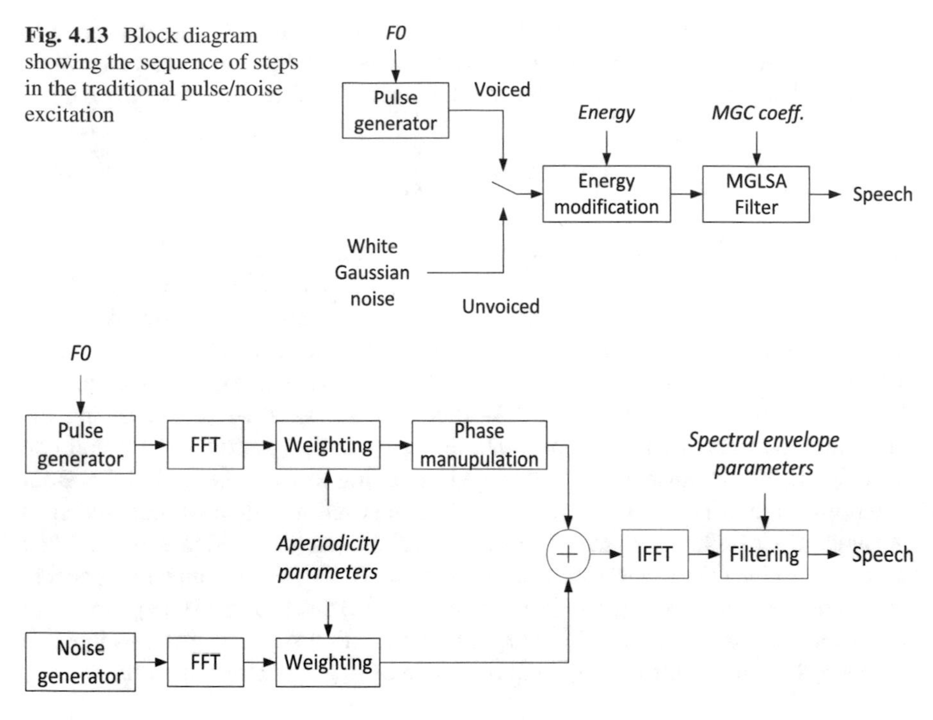

Fig. 4.13 Block diagram showing the sequence of steps in the traditional pulse/noise excitation

Fig. 4.14 Block diagram showing the sequence of steps in the STRAIGHT vocoder

approach for source modeling. In STRAIGHT-HTS, mel-cepstral coefficients are extracted from F0-adaptive smoothed spectral envelope. The aperiodicity measures are computed based on the ratio of the lower and upper smoothed spectral envelopes and averaged across five frequency sub-bands (0–1, 1–2, 2–4, 4–6, and 6–8 kHz). Figure 4.14 shows the block diagram indicating the sequence of steps in the STRAIGHT vocoder. The excitation signal consists of a sequence of impulses and noise components weighted by band-pass filtered aperiodicity parameters. At each frequency bin, the aperiodicity parameter is converted to the weight for a noise signal by using a sigmoid mapping function [25]. As suggested in [26], the phase of the impulses is manipulated so as to reduce buzziness. Pitch-synchronous overlap-add method is used to generate the excitation signal, which is then passed through a synthesis filter to generate the speech signal.

3. **DSM-HTS [14]:** DSM-HTS is one of the recently proposed popular approaches which models the pitch-synchronous residual frames based on the deterministic plus stochastic model. In DSM-HTS, the spectrum of the pitch-synchronous residual frame is divided into two bands delimited by a maximum voiced frequency. Both components are extracted from an analysis performed on a speaker-dependent dataset of pitch-synchronous residual frames. The deterministic and stochastic components model the low- and high-frequency contents of the residual frame, respectively. The first eigenvector obtained from the principal

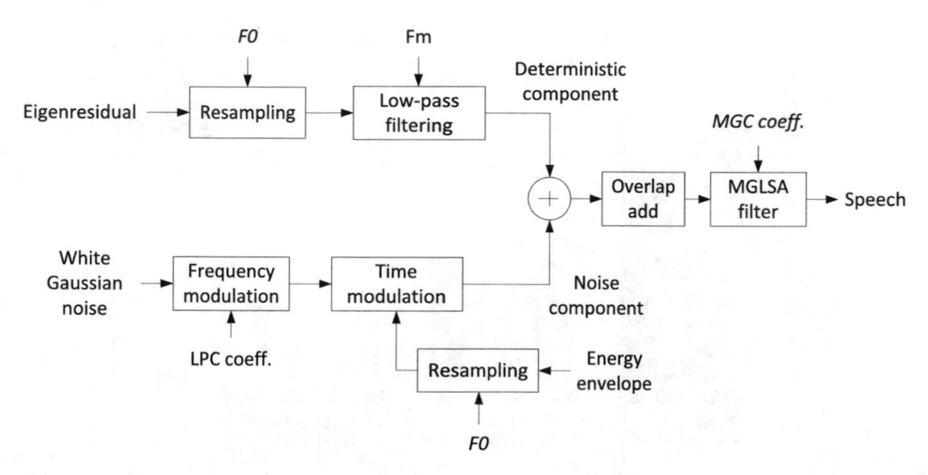

Fig. 4.15 Block diagram showing the sequence of steps in the DSM vocoder

component analysis of the pitch-synchronous residual frames is considered to represent the deterministic component. The stochastic component is obtained by convolving an autoregressive model (represented as LPC coefficients) with white Gaussian noise, and the time structure of the resulting signal is controlled by an energy envelope. The autoregressive model is estimated from the average spectral envelope of high-frequency contents. The energy envelope is determined as the average Hilbert envelope of high-frequency contents. During synthesis, the excitation signal is constructed by modifying the deterministic component and the energy envelope of noise component according to the generated pitch. The block diagram showing the sequence of steps in the DSM vocoder is given in Fig. 4.15.

Before evaluation, the energies of the speech signals synthesized from different source modeling methods are normalized to the same level. CMOS scores with 95% confidence intervals and preference scores are provided in Figs. 4.16 and 4.17, respectively. On comparing the proposed method with the Pulse-HTS, it can be observed that the CMOS scores are varying between 1.1 and 1.5 and more than 60% of the subjects preferred the proposed method for both male and female speakers. This indicates that the quality of the speech synthesized by the proposed method is clearly better than the speech synthesized by the Pulse-HTS. The subjects noticed that the speech synthesized from the Pulse-HTS is artificial and unnatural. The excitation signal generated using the combination of deterministic and noise components is much better than the sequence of pulses.

On comparison of the proposed method with the STRAIGHT-HTS, it can be observed that the CMOS scores vary between 0.4 and 0.7, and the subjects preferred the proposed method for about 40% of the cases and preferred the STRAIGHT-HTS for about 28% of the cases. Both measures indicate that the proposed method is better than the STRAIGHT-HTS. The STRAIGHT vocoder uses the mixed

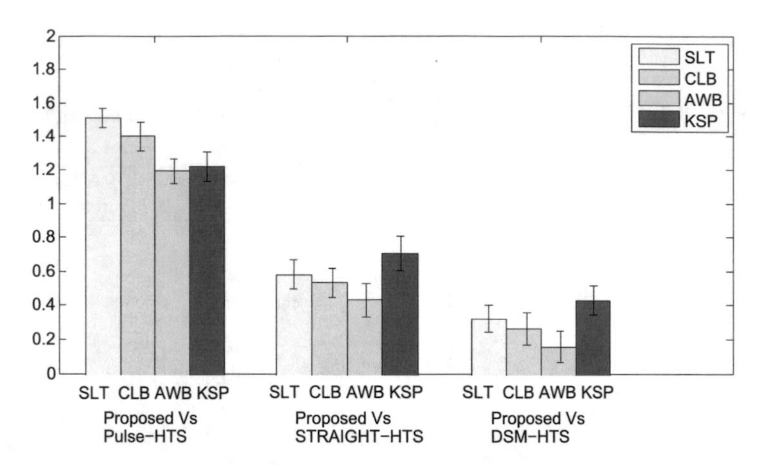

Fig. 4.16 CMOS scores with 95% confidence intervals obtained by comparing the proposed method with three existing methods

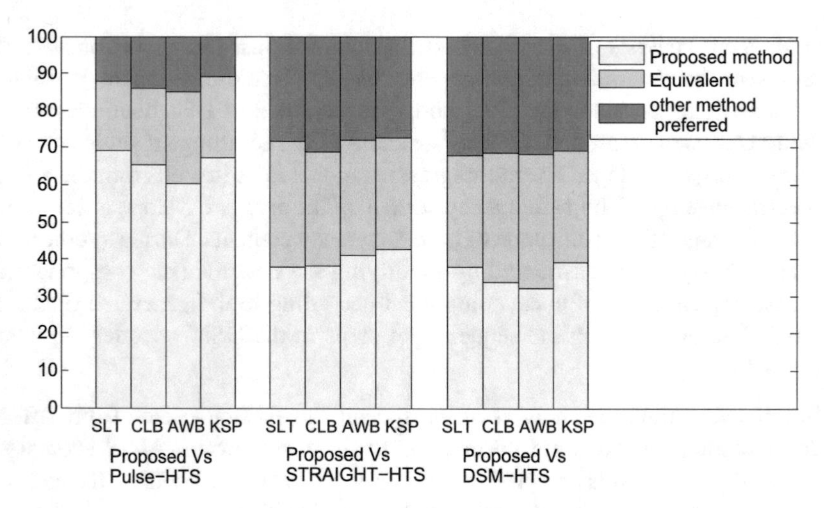

Fig. 4.17 Preference scores obtained by comparing the proposed method with three existing methods

excitation parameters to model and generate the voice source signal. In the proposed method, the deterministic component or the segment of residual signal around GCI, which is important for the perception of speech, is accurately represented. Hence, the generated excitation signal is close to the natural source signal.

Regarding the comparison of the proposed method with the DSM-HTS, it can be observed that the CMOS scores are varying in the range of 0.1–0.4, and the subjects preferred the proposed method for about 35% of the cases and preferred the DSM-HTS for about 30% of the cases. Both measures show that the proposed method is slightly better than the DSM-HTS. The DSM-HTS utilizes the single instance of the deterministic (first eigenvector) and noise component (the average spectral and amplitude envelopes modulated white Gaussian noise) to generate the excitation signal. This kind of modeling cannot incorporate the natural variations of the excitation signal from one pitch-synchronous residual frame to another. In the proposed method, the deterministic and noise components are generated for every cycle of the pitch-synchronous residual frame. This results in the incorporation of characteristics of the real voice source in the excitation signal.

The synthesized speech samples of the proposed parametric source modeling approach and the three existing source modeling methods are made available online at http://www.sit.iitkgp.ernet.in/~ksrao/narendra-phd-demo/narendra_hsm.html.

To analyze the effectiveness of the excitation models without any influence from statistical models, the natural excitation signal is modeled using four source models, namely, (1) pulse, (2) STRAIGHT, (3) DSM, and (4) proposed parametric method. Using the natural spectrum, F0 and excitation signals constructed from four source models, the speech signals are synthesized. Figure 4.18 shows the natural speech, excitation signal, synthesized speech, and corresponding excitation signals constructed from four source modeling methods. The excitation signal generated by the pulse-based source model (Fig. 4.18d) is having non-zero amplitude values only at GCIs. This kind of excitation signal is an imprecise approximation of natural excitation signal (Fig. 4.18b). From the excitation signal generated from STRAIGHT source model (Fig. 4.18f), it can be observed that in addition to non-zero amplitude values at GCIs, small amount noise is also present around GCIs. On comparing this signal with the natural excitation signal, significant differences can be observed in the waveform shapes of each pitch cycle. The excitation signal constructed from DSM-based source model (Fig. 4.18h) is closer to natural excitation compared to pulse and STRAIGHT source models. But, as excitation signal is generated by resampling single instance of deterministic and noise components, cycle to cycle variations present in the natural excitation signal are not observed in the excitation signal of the DSM-based source model. The excitation signal constructed from the proposed method (Fig. 4.18j) is very much close to the natural excitation signal (Fig. 4.18b) and better compared to three other source models. The speech waveform produced by the proposed source model (Fig. 4.18i) is also very close to the natural speech waveform (Fig. 4.18a).

4.3 Summary

This chapter proposed two parametric methods for modeling the excitation signal. The first method parameterized the pitch-synchronous residual frames of the excitation signal by using 30 PCA coefficients. During synthesis, the residual frame is reconstructed from 30 PCA coefficients. In the second method, the pitch-synchronous residual frames of the excitation signal are modeled as deterministic and noise components. Initially, the analysis of characteristics of the residual frames around GCI is performed using PCA. Based on the analysis, the segment of the residual frame around GCI is considered as the deterministic component, and

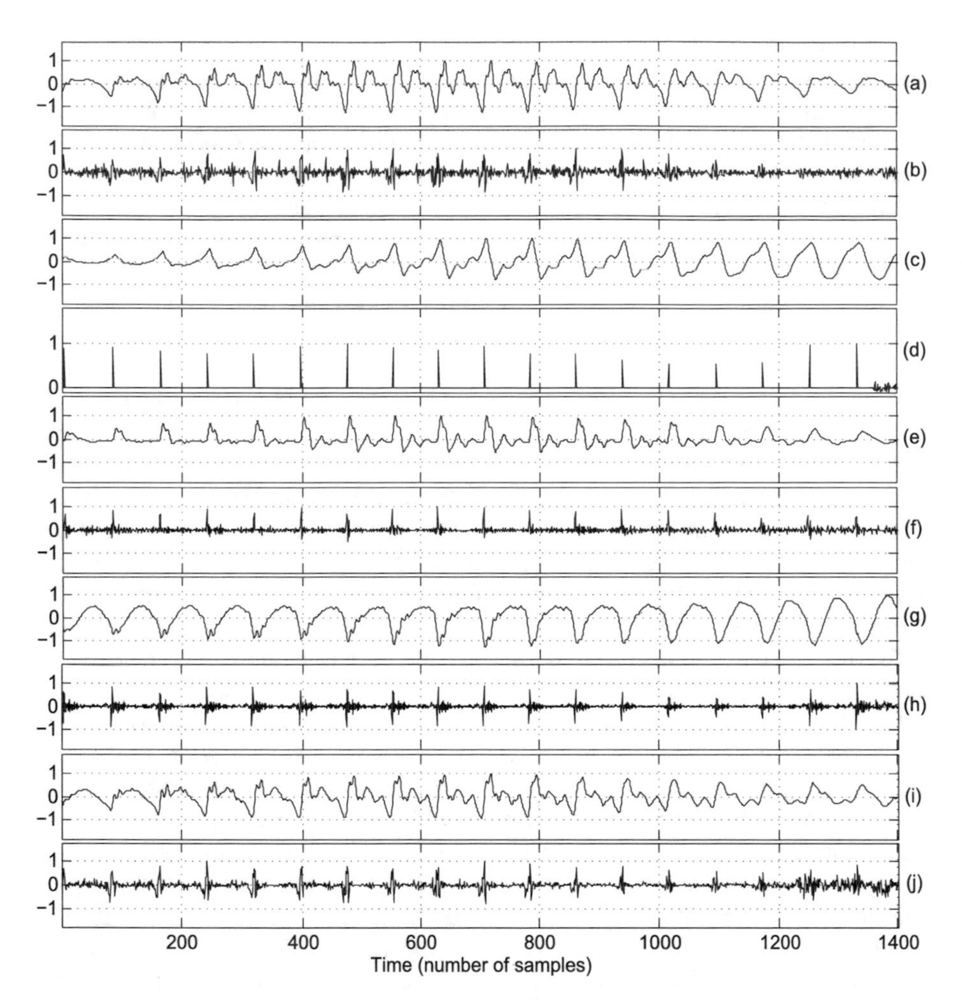

Fig. 4.18 (**a**) Natural speech, (**b**) excitation signal, (**c**) speech synthesized by pulse-based source model, (**d**) excitation signal constructed by pulse-based source model, (**e**) speech synthesized by STRAIGHT source model, (**f**) excitation signal constructed by STRAIGHT source model, (**g**) speech synthesized by DSM-based source model, (**h**) excitation signal constructed by DSM-based source model, (**i**) speech synthesized by proposed source model, and (**j**) excitation signal constructed by proposed parametric source model

the remaining part of the residual frame is considered as the noise component. The deterministic components are accurately modeled using PCA coefficients. The noise components are parameterized in terms of spectral and amplitude envelopes. During synthesis, the deterministic and noise components are reconstructed from the parameters generated by HMMs. The subjective evaluation results indicated that the quality of the proposed method is considerably better compared to three existing excitation modeling methods.

References

1. K. Tokuda, T. Kobayashi, T. Masuko, S. Imai, Mel-generalized cepstral analysis a unified approach to speech spectral estimation, in *Proceedings of International Conference on Spoken Language Processing (ICSLP)* (1994), pp. 1043–1046
2. H. Zen, T. Toda, K. Tokuda, The Nitech-NAIST HMM-based speech synthesis system for the Blizzard Challenge 2006. IEICE Trans. Inf. Syst. **E91-D**(6), 1764–1773 (2008)
3. K.S.R. Murty, B. Yegnanarayana, Epoch extraction from speech signals. IEEE Trans. Audio Speech Lang. Process. **16**(8), 1602–1613 (2008)
4. T. Drugman, G. Wilfart, T. Dutoit, Eigenresiduals for improved parametric speech synthesis, in *Proceedings of European Signal Processing Conference (EUSIPCO)* (2009), pp. 2177–2180
5. B. Sch, A. Smola, Nonlinear component analysis as a kernel eigenvalue problem. Neural Comput. **10**(5), 1299–1319 (1998)
6. J. Gudnason, M.R.P. Thomas, D.P. Ellis, P.A. Naylor, Data-driven voice source waveform analysis and synthesis. Speech Commun. **54**(2), 199–211 (2012)
7. I. Jolliffe, *Principal Component Analysis* (Wiley, Hoboken, 2002)
8. CMU ARCTIC speech synthesis databases [Online]. http://festvox.org/cmu_arctic/
9. K.S. Rao, B. Yegnanarayana, Prosody modification using instants of significant excitation. IEEE Trans. Audio Speech Lang. Process. **14**(3), 972–980 (2006)
10. J.P. Cabral, Uniform concatenative excitation model for synthesising speech without voiced/unvoiced classification, in *Proceedings of Interspeech* (2013), pp. 1082–1086
11. N. Adiga, S.R.M. Prasanna, Significance of instants of significant excitation for source modeling, in *Proceedings of Interspeech* (2013), pp. 1677–1681
12. E. Yumoto, W. Gould, T. Baer, Harmonics-to-noise ratio as an index of the degree of hoarseness. J. Acoust. Soc. Am. **71**(6), 1544–1550 (1982)
13. T. Drugman, A. Moinet, T. Dutoit, G. Wilfart, Using a pitch-synchrounous residual codebook for hybrid HMM/frame selection speech synthesis, in *Proceedings of International Conference on Acoustics, Speech and Signal Processing, (ICASSP)* (2009), pp. 3793–3796
14. T. Drugman, T. Dutoit, The deterministic plus stochastic model of the residual signal and its applications. IEEE Trans. Audio Speech Lang. Process. **20**(3), 968–981 (2012)
15. T. Raitio, A. Suni, J. Yamagishi, H. Pulakka, J. Nurminen, M. Vainio, P. Alku, HMM-based speech synthesis utilizing glottal inverse filtering. IEEE Trans. Audio Speech Lang. Process. **19**(1), 153–165 (2011)
16. T. Drugman, G. Wilfart, T. Dutoit, A deterministic plus stochastic model of the residual signal for improved parametric speech synthesis, in *Proceedings of Interspeech* (2009), pp. 1779–1782
17. J. Cabral, S. Renals, J. Yamagishi, K. Richmond, HMM-based speech synthesiser using the LF-model of the glottal source, in *Proceedings of IEEE International Conference on Acoustics, Speech and Signal Processing (ICASSP)* (2011), pp. 4704–4707
18. X. Huang, A. Acero, H.W. Hon, *Spoken Language Processing: A Guide to Theory, Algorithm and System Development* (Prentice Hall, Upper Saddle River, 2001)
19. F. Soong, B.-H. Juang, Line spectrum pair (LSP) and speech data compression, in *Proceedings of International Conference on Audio, Speech and Signal Processing (ICASSP)* (1984), pp. 37–40
20. K. Paliwal, W. Kleijn, Quantization of LPC parameters, in *Speech Coding and Synthesis* (Elsevier, Amsterdam, 1995)
21. Z. Ling, Y. Wu, Y. Wang, L. Qin, R. Wang, USTC system for Blizzard Challenge 2006: an improved HMM-based speech synthesis method, in *Blizzard Challenge Workshop* (2006)
22. Y. Pantazis, Y. Stylianou, Improving the modeling of the noise part in the harmonic plus noise model of speech, in *Proceedings of IEEE International Conference on Acoustics, Speech and Signal Processing (ICASSP)* (2008), pp. 4609–4612

23. K. Tokuda, T. Yoshimura, T. Masuko, T. Kobayashi, T. Kitamura, Speech parameter generation algorithms for HMM-based speech synthesis, in *Proceedings of International Conference on Acoustics, Speech, and Signal Processing (ICASSP)* (2000), pp. 1315–1318
24. H. Zen, T. Toda, M. Nakamura, K. Tokuda, Details of Nitech HMM-based speech synthesis system for the Blizzard Challenge 2005. IEICE Trans. Inf. Syst. **E90-D**(1), 325–333 (2007)
25. Y. Ohtani, T. Toda, H. Saruwatari, K. Shikano, Maximum likelihood voice conversion based on GMM with STRAIGHT, in *Proceedings of Interspeech* (2006), pp. 2266–2269
26. H. Kawahara, I. Masuda-Katsuse, A. de Cheveigne, Restructuring speech representations using a pitch-adaptive time-frequency smoothing and an instantaneous-frequency-based F0 extraction: possible role of a repetitive structure in sounds. Speech Commun. **27**(3–4), 187–207 (1999)

Chapter 5
Hybrid Approach of Modeling the Source Signal

5.1 Optimal Residual Frame-Based Hybrid Source Modeling Method

The proposed hybrid source modeling method models the excitation signal by utilizing the optimal residual frames extracted from every phone. Flowchart indicating the sequence of steps in the proposed hybrid source modeling method based on the optimal residual frame is shown in Fig. 5.1. Initially, energy is extracted from every frame of the excitation signal. Then, the pitch-synchronous residual frames are extracted from the excitation signal. The detailed procedure for extracting the pitch-synchronous residual frames is given in Sect. 4.1.1. From the pitch-synchronous residual frames of every phone, the optimal residual frames are extracted (discussed in Sect. 5.1.1). The optimal residual frame closely represents all the pitch-synchronous residual frames of a phone. The optimal residual frames extracted from all phone present in the database are clustered in the form of a decision tree (discussed in Sect. 5.1.2). During synthesis, for every input phone, a suitable optimal residual frame is selected from the leaf of the decision tree based on target and concatenation costs (explained in Sect. 5.1.3). Using single optimal frame, the excitation signal of a phone is constructed. The optimal residual frame is resampled to twice the target pitch period (which is generated by the HMM). The pitch modified optimal residual frames are overlap-added to generate the excitation signal of the phone.

5.1.1 Computation of Optimal Residual Frame for a Phone

The number of pitch-synchronous residual frames extracted from the excitation signal of a phone varies from one phone to another. On observing the pitch-synchronous residual frames of different phones, it is observed that the shapes of

K. S. Rao, N. P. Narendra, *Source Modeling Techniques for Quality Enhancement in Statistical Parametric Speech Synthesis*, SpringerBriefs in Speech Technology,
https://doi.org/10.1007/978-3-030-02759-9_5

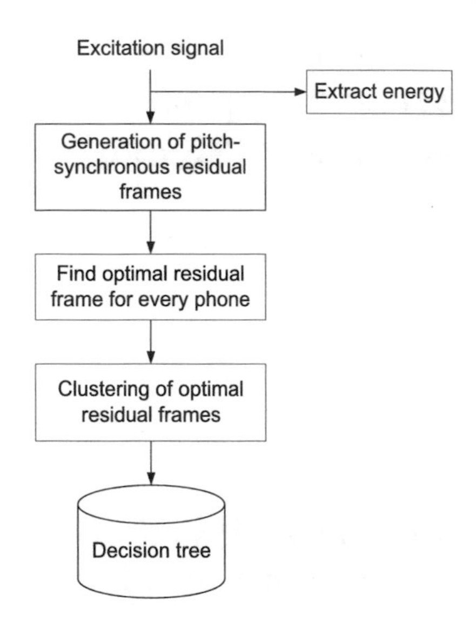

Fig. 5.1 Flowchart indicating the sequence of steps in the proposed hybrid source modeling method based on optimal residual frame

the residual frames are very similar within the phone, and between different phones the residual frames have different shapes. The adjacent pitch-synchronous residual frames of a phone exhibit strong correlation compared to the residual frames present in other phones. Hence, instead of considering all residual frames, one optimal residual frame is selected which closely represents the candidate of all residual frames of a phone. This optimal residual frame is used for representing the excitation signal of a phone. To select the optimal frame, every residual frame present in a phone is considered, and its acoustic distance with all other residual frames present in a phone is computed. The residual frame which is close to all the residual frames in the phone is considered as the optimal residual frame for the phone. To compute the acoustic distance between two residual frames, a set of acoustic features need to be computed. The acoustic features used for representing the residual frame are as follows:

1. Energies in six parts of the residual frame (six dimensions)
2. Ratio of maximum to minimum peak amplitude of the residual frame (one dimension)
3. Pulse spread of the residual frame (one dimension)
4. Linear predictor coefficients (ten dimensions)

First three features, namely, energies in six parts, ratio of maximum to minimum peak amplitude, and pulse spread of residual frame, try to capture the time-domain variations of residual frame. The single instance of residual frame indicating different features is shown in Fig. 5.2. The residual frame is divided into six equal parts (S1–S6). In every part of the residual frame, energy is computed. Computation of energy in this way provides information about the distribution of energy in the residual frame. The residual frames of vowels have energy distributed mostly in the middle portion (S3 and S4) of the frame. The residual frames of voiced fricatives

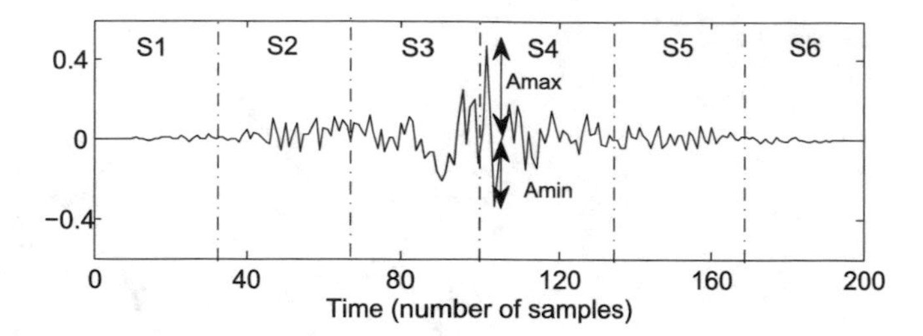

Fig. 5.2 Single instance of the residual frame indicating different features considered for computing the acoustic distance

have a significant amount of energy in S2 and S5 regions. The ratio of maximum (Amax) to minimum (Amin) peak indicates the orientation of the residual frame, whether the residual frame is oriented toward the positive or negative direction. The pulse spread is the time interval (in the number of samples) where the peak of Hilbert envelope of the residual frame drops to certain threshold α (the value of α is chosen to be 20%). Ten-dimensional LPCs represent the spectral envelope of the residual frame.

Using the proposed features, the acoustic distance is computed between every residual frame in a phone with all other residual frames. The acoustic distance is the Mahalanobis distance of features of two residual frames. Mean acoustic distance is calculated between a given residual frame and all other residual frames. This mean acoustic distance provides the impurity of given residual frame with other residual frames of a phone. Similarly, the impurities of all residual frames of a phone are computed. The residual frame which has the lowest impurity is considered as the optimal residual frame of a particular phone.

5.1.2 Clustering the Optimal Residual Frames

During synthesis, the optimal residual frames of all phones are available for generating the excitation signal. The optimal residual frame which best matches with the given target phone specification is selected from the database of optimal residual frames. To accomplish this, the optimal residual frames are grouped into a number of clusters. All the units present in the cluster are acoustically close to each other. For the given target unit specification, the appropriate cluster is selected which offers a small set of candidate optimal residual frames. The suitable residual frame is selected from the cluster based on target and concatenation costs.

To cluster the optimal residual frames, the acoustic distance between optimal residual frames needs to be computed. To compute acoustic distance, the optimal residual frames are represented by a set of features as described in Sect. 5.1.1. The acoustic distance is the Mahalanobis distance of features of two optimal residual frames. Using this acoustic measure, the acoustic impurity of the cluster of units

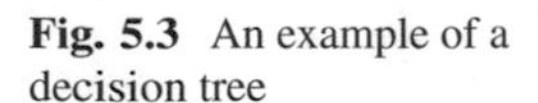
Fig. 5.3 An example of a decision tree

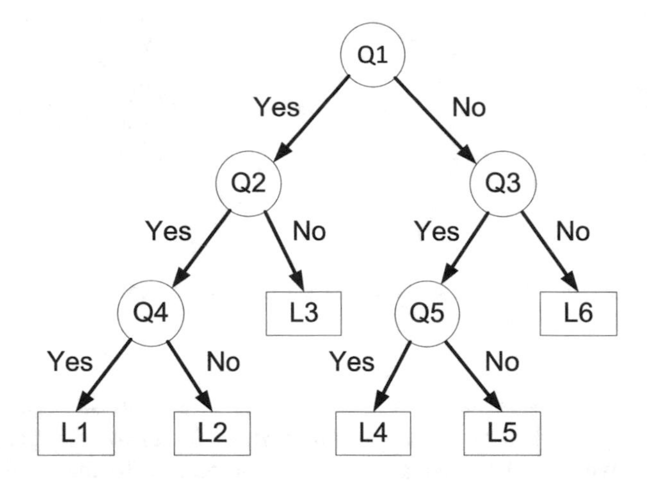

is computed which is the mean acoustic distance of all the units present in the cluster. A decision tree is constructed by splitting the cluster of optimal residual frames into subclusters based on the questions related to positional and contextual features of a phone. The questions that best minimize the acoustic impurity of subclusters are asked at each node of the decision tree. The question set consists of a standard list of 53 positional and contextual features provided in basic HTS toolkit [1]. Classification and regression tree (CART) method [2] is used to build the decision tree. The minimum number of units present in the cluster is specified to be 10. By considering the optimal residual frames of all phones present in the database, a single decision tree is constructed. An example of such a decision tree is shown in Fig. 5.3. In the figure, Q1–Q5 are the questions asked at the nodes of the decision tree, and L1–L6 are the leaves containing the cluster of residual frames.

In this hybrid source model, the excitation parameters consist of F_0 with voicing decision and frame-wise energy, and spectral parameters include MGC coefficients (34th order with $\alpha = 0.42$, $Fs = 16\,\text{kHz}$, and $\gamma = -1/3$). Both MGC coefficients and energy values are modeled by using CD-HMMs, and F_0 patterns are modeled by using MSD-HMMs.

5.1.3 Speech Synthesis Using the Proposed Optimal Residual Frame-Based Hybrid Source Model

During synthesis, the MGC coefficients, F0 including voicing decision and energy, are generated by HMMs. The block diagram showing different stages of synthesis is presented in Fig. 5.4. In the figure, the parameters generated by the HMMs are shown in italics. The positional and contextual features are extracted from each phone present in the input text. Based on these features, the cluster of units present in the leaf of the decision tree is selected by answering the questions at each node.

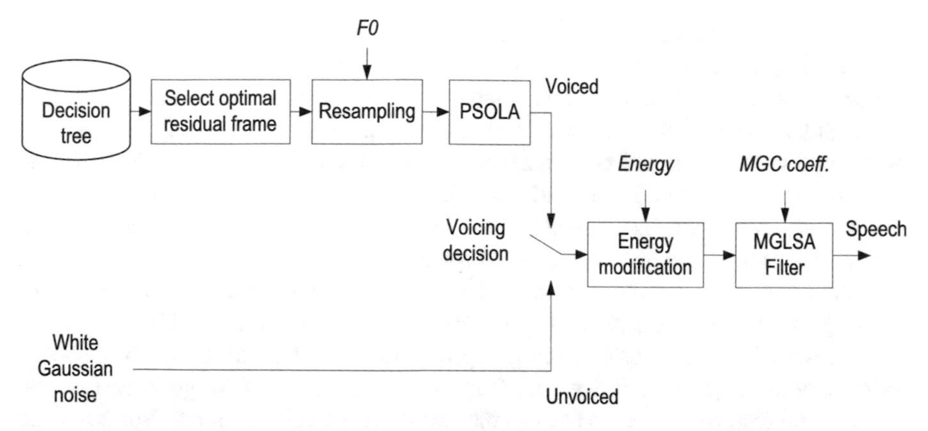

Fig. 5.4 Block diagram showing different synthesis stages in the proposed optimal residual frame-based hybrid source model. The parameters generated by the HMMs are shown in italics

To select the most appropriate unit from the cluster of candidate units, target and concatenation costs are computed. The target cost gives the estimate of the closeness of a candidate unit in the cluster with the target unit. Features used as the subcosts of target cost are position of phrase (PP), position of word (PW), position of syllable (PS), phone identity (PI), previous phone identity (PPI), and next phone identity (NPI). Based on the position of a syllable in the word, PS is categorized as begin, middle, and end. Similarly based on the position of a word in the phrase and the position of a phrase in the sentence, PP and PW are also classified as begin, middle, and end. If the target unit specification matches with that of the candidate unit, then subcost is assigned as zero; otherwise, it is one. The target cost ($C^t(t_i, u_i)$) is given by

$$C^t(t_i, u_i) = \sum_{j=1}^{p} w_j^t C_j^t(t_i, u_i) \tag{5.1}$$

where t_i is the target unit, u_i is a candidate unit, p is the number of subcosts used to find the target cost, C_j^t is the jth target subcost, and w_j^t is the weight given to the jth target subcost.

The residual frames present in adjacent phones cannot vary drastically. Hence, in order to ensure smooth variation at adjacent phones, concatenation cost is computed. The acoustic features used for describing the residual frames are employed as the subcosts of concatenation cost. The concatenation cost is computed as the weighted sum of difference between the acoustic features of two candidate units. The concatenation cost ($C^c(u_{i-1}, u_i)$) is given by

$$C^c(u_{i-1}, u_i) = \sum_{j=1}^{q} w_j^c C_j^c(u_{i-1}, u_i) \tag{5.2}$$

where C_j^c is jth concatenation subcost, q is the number of subcosts used to find the concatenation cost, and w_j^c is the weight for jth concatenation subcost. The sequence of candidate units which has the lowest sum of target and concatenation costs are chosen as the best sequence of residual frames of the utterance. The weights for the target and concatenation costs are determined manually based on informal listening tests. In most of the unit selection synthesis methods, the target and the concatenation costs are determined based on informal listening tests [3, 4]. Hence, the same approach was adopted for determining the weights for target and concatenation costs. The chosen optimal residual frames are resampled according to the target pitch contour generated by the HMM. The excitation signal is constructed by pitch-synchronous overlap-adding of pitch modified residual frames. The energy of excitation signal is modified according to the energy measure generated by the HMM. The resulting signal is used as the excitation of voiced frames. For unvoiced frames, the energy of white Gaussian noise is modified according to the generated energy measure. The generated excitation is finally given as input to MGLSA filter to produce the speech.

5.1.4 Evaluation

As followed in previous chapters, the proposed method is evaluated using four speakers (SLT, CLB, AWB, and KSP) from CMU Arctic speech database [5]. Subjective evaluation is performed using two measures, namely, CMOS and preference tests. Depending on the number of units present in the clusters of the decision tree, two versions of the proposed method, namely, full-cluster and pruned-cluster, are considered for the evaluation. In the full-cluster, all units present in the leaves of the decision tree are utilized for the generation of the excitation signal. The pruned-cluster contains only one optimal residual frame at every leaf of the decision tree. The pruned-cluster is obtained by removing all units from the cluster that are farthest from the cluster center (mean acoustic difference of all the units present in the cluster) and retaining only one unit which is close to the cluster center. As pruned-cluster contains only one unit at every leaf, the optimal residual frame is chosen directly from the leaf without computing any target and concatenation costs. Two versions of the proposed method are compared with Pulse-HTS.

The CMOS with 95% confidence intervals and the preference scores are provided in Figs. 5.5 and 5.6, respectively. From the figure, it can be observed that the CMOS scores vary between 1 and 1.4 for both male and female speakers. The preference scores provided in Fig. 5.6 show that the subjects preferred the proposed method for about 60% of cases. This indicates that the quality of the speech synthesized by both full-cluster and pruned-cluster versions of the proposed method is clearly superior compared to the pulse-HTS. The CMOS and preference scores of the full-cluster version are slightly more compared to the pruned-cluster. This is because the full-cluster version offers more variation of the residual frames than pruned cluster. But

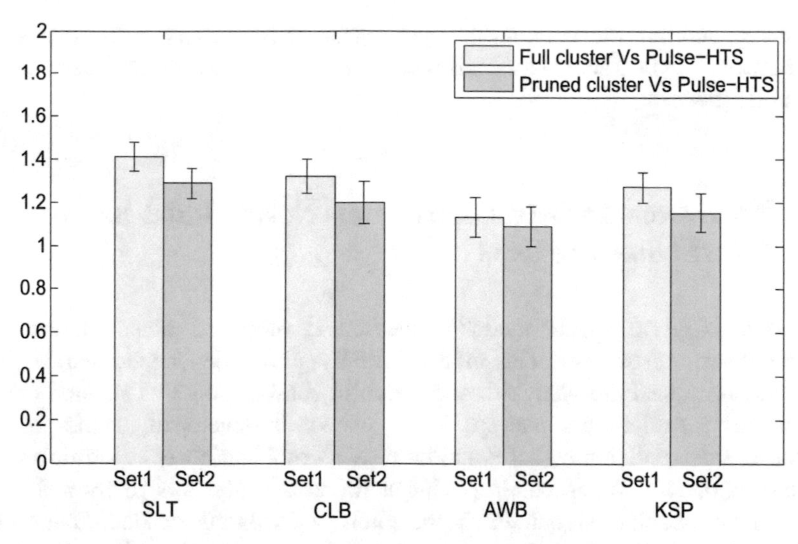

Fig. 5.5 CMOS scores with 95% confidence intervals obtained by comparing the proposed method with the Pulse-HTS

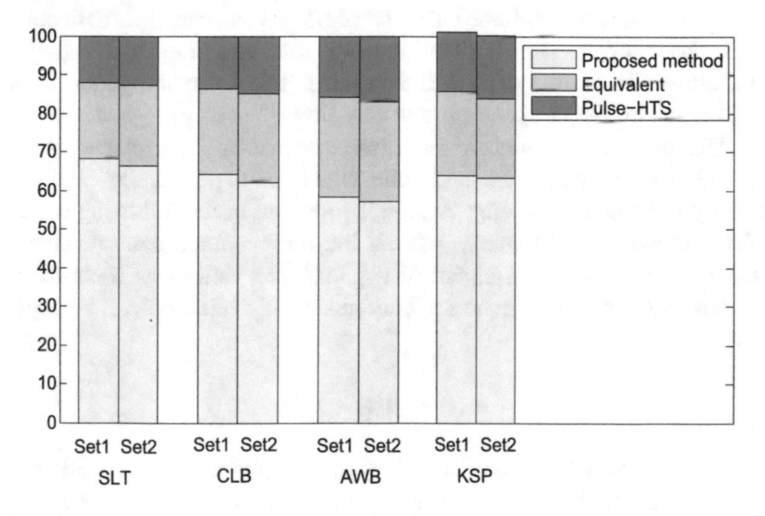

Fig. 5.6 Preference scores obtained by comparing the proposed method with the Pulse-HTS

the advantage of the pruned-cluster version over the full-cluster version is that it requires less memory to store the small number of units (statistics of memory usage is discussed in detail in Sect. 5.2.6).

The proposed hybrid source model used single optimal residual frame to construct the excitation signal of a phone. Even though the optimal residual frame closely represents all the residual frames in a phone, there exists slow variation in every cycle of the residual frames of natural excitation signal. To generate the excitation signal close to natural quality, every residual frame should be represented accurately, and variations across the cycles need to be preserved. To address this requirement, a hybrid source model is proposed based on time-domain deterministic plus noise model.

5.2 Time-Domain Deterministic Plus Noise Model-Based Hybrid Source Model

The second hybrid source modeling method is proposed based on the time-domain deterministic plus noise model. Initially, phone-dependent characteristics of excitation signal are analyzed, and a hybrid source modeling method capable of generating excitation signal specific to phones is developed. In the proposed approach, the excitation signal is modeled as a combination of deterministic and noise components. Time-domain pitch-synchronous analysis is performed on the excitation or residual signal. From the pitch-synchronous residual frames of a phone, the deterministic and noise components are estimated. The deterministic components of all phones are systematically arranged in the form of a decision tree. The spectrum and amplitude envelope of noise components are modeled using hidden Markov models (HMMs). During synthesis, the suitable deterministic component is chosen from the leaf of a decision tree. The noise component is obtained after imposing the target spectrum and amplitude envelopes generated from the HMMs. The sum of deterministic and noise components are pitch-synchronously overlap-added to construct the excitation signal of a phone. Description of the proposed hybrid source modeling method is provided in the following section.

In the proposed hybrid source model, the pitch-synchronous residual frames extracted from the excitation signal of a phone are viewed as a combination of deterministic and noise components. This model of excitation can be represented as follows:

$$e_i(t) = p(t) + r_i(t) \tag{5.3}$$

where $e_i(t)$ is the excitation signal of ith cycle of a phone, $p(t)$ is the deterministic component, and $r_i(t)$ is the noise component of the excitation signal $e_i(t)$. In this work, it is assumed that the deterministic component is constant for all pitch cycles of a phone and the noise component varies for every pitch cycle of a phone. The deterministic component $p(t)$ is computed as an ensemble average of individual pitch cycles of a phone, which is given by

$$p(t) = \frac{\sum_{i=1}^{N} e_i(t)}{N} \tag{5.4}$$

where N is the number of pitch cycles present in a phone. The duration of all pitch periods in a phone is not exactly equal, and there exists variation from one cycle to another. In the presence of pitch perturbations, the computed deterministic

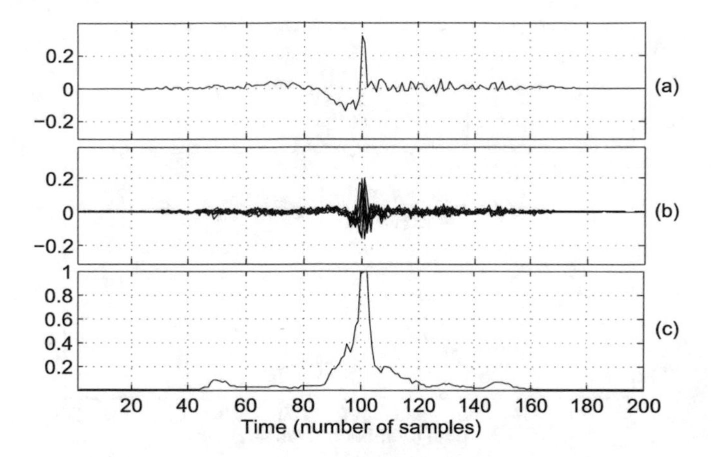

Fig. 5.7 (**a**) Deterministic component, (**b**) noise components superimposed one over the other, and (**c**) variance of noise components, for the phone /aa/

component is not accurate as individual pitch cycles are not aligned properly. To have a better estimate of the deterministic component, every pitch cycle is normalized in time to a maximum pitch period (F_0^*) as explained in Sect. 4.1.1. The noise component of each pitch cycle is computed by subtracting the deterministic component from individual pitch cycles. The noise component $r_i(t)$ computed for the ith pitch cycle is given by

$$r_i(t) = e_i(t) - p(t) \tag{5.5}$$

The deterministic and noise components are unique for every phone. Analysis of the variation of deterministic and noise components for different phonetic classes is performed. From the noise components estimated from every pitch cycle of a phone, variance of noise at each sample value is computed. Figures 5.7 and 5.8 provide the deterministic component, superimposed segments of noise components, and variance of noise components for single instance of phones /aa/ and /z/, respectively. From Fig. 5.7c, it can be observed that the variance of noise components of /aa/ is concentrated around GCI, whereas in Fig. 5.8c, the variance of noise components of /z/ is spread out more in time compared to the variance of noise components of /aa/. The shapes of the variance of noise components change for different phonetic classes. Similar kind of variations can be observed for the variance of deterministic components of various phonetic classes.

To quantify the changes in the shape of the variance of noise and deterministic components for different phonetic classes, the sharpness of the variance is computed. The sharpness of the variance is calculated similar to the sharpness of the peaks in the Hilbert envelope of the LP residual as proposed in [6]. The sharpness is given by $\eta = \sigma/\mu$, where μ and σ denote mean and standard deviation of the variance of noise or deterministic components, respectively. Ideally, if the variance

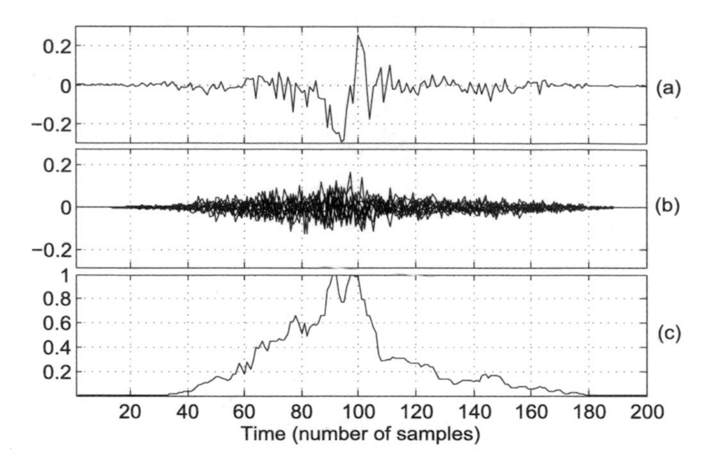

Fig. 5.8 (**a**) Deterministic component, (**b**) noise components superimposed one over the other, and (**c**) variance of noise components, for the phone /z/

of noise or deterministic components of length N has an impulse (in the discrete-time domain) at GCI, then η has a maximum value of $\sqrt{N}$, and if the variance has a constant value throughout its length, then η is zero. 100 instances of every phonetic class are considered for the analysis. The noise components are computed from every instance of a phonetic class. The noise components obtained from 100 instances of a phonetic class are collected, and these are considered for calculating the variance of noise components. From 100 instances of a phonetic class, 100 deterministic components are obtained, and these are considered for computing the variance of deterministic components. From the variances of deterministic and noise components, the sharpness of the variance of noise (η_N) and deterministic components (η_D) are computed. η_N and η_D determined for different phonetic classes for speaker SLT from the CMU ARCTIC database [5] are shown in Table 5.1. From Table 5.1, the following observations can be drawn:

1. The values of η_N and η_D are varying proportionately for all phonetic classes. Whenever η_N is high (or low) for a phonetic class, correspondingly η_D is also high (or low).
2. Variations in η_N and η_D can be observed between broad phonetic classes such as vowels, nasals, semivowels, and voiced fricatives. But within the broad phonetic class, the variation in the values of η_N and η_D are not significant.
3. For vowels, η_N and η_D are more compared to other phones. Within the vowels, /aa/, /ae/, and /oy/ have the highest values. This indicates that the variance of noise and deterministic components are concentrated sharply around the GCI for these vowels.
4. Semivowels /y/, /w/, and /r/ have little less sharpness values compared to vowels but higher than nasals, voiced fricatives, and voiced plosives.
5. Voiced fricatives /dh/ and /z/ have the lowest η_N and η_D values. This indicates that the variance of noise and deterministic components are spread throughout its

Table 5.1 Sharpness of variance of noise and deterministic components for different phonetic classes

Phone	η_N	η_D
aa	1.811	1.723
ae	1.885	1.680
ah	1.182	1.272
ao	1.063	1.335
aw	0.979	1.146
ax	1.117	0.832
ay	1.067	1.251
eh	1.161	0.989
er	1.188	1.094
ey	1.123	1.247
ih	1.298	1.030
iy	0.963	1.199
ow	1.183	1.362
oy	1.577	1.887
uh	1.081	1.133
uw	1.066	1.341
d	0.941	0.975
dh	0.625	0.658
g	0.936	0.993
l	0.997	1.164
m	0.756	0.864
n	0.771	0.822
ng	0.824	0.914
r	1.051	1.126
v	0.667	0.715
w	1.069	1.132
y	1.102	1.156
z	0.590	0.606

length. The reason for the higher spread of variance is due to breathy phonation of voiced fricatives.

These studies imply that the deterministic and noise components have phone-dependent characteristics. Using the appropriate deterministic and noise components of a phone results in the generation of proper excitation signal close to the natural quality.

The proposed hybrid source modeling method based on time-domain deterministic plus noise model is presented in Fig. 5.9. First, energy is extracted from every frame of the excitation signal. Then, the pitch-synchronous analysis is performed on the excitation signal leading to a set of residual frames that are synchronous with the GCI and whose length is set to two pitch periods (discussed in Sect. 4.1.1). The time-domain deterministic plus noise modeling is performed on the set of pitch-synchronous residual frames of a phone. Single deterministic component

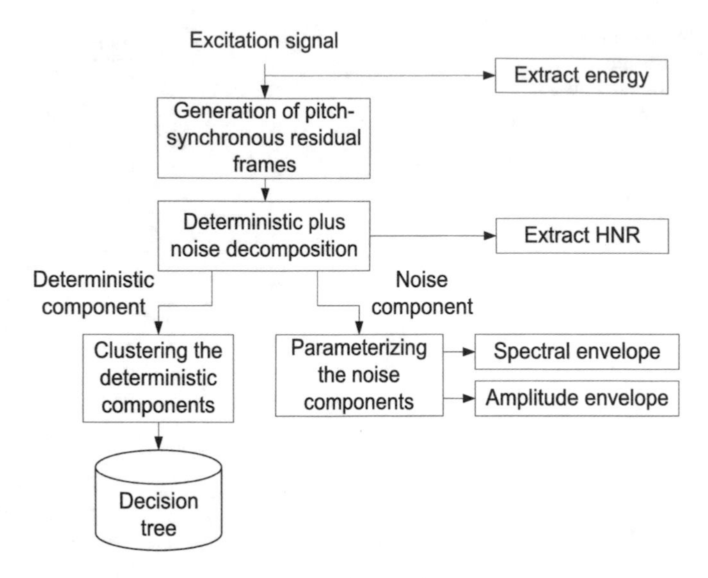

Fig. 5.9 Flowchart indicating the sequence of steps in the proposed hybrid source modeling method based on time-domain deterministic plus noise model

and multiple noise components (i.e., the number of noise components is equal to the number of residual frames of the phone) of the phone are estimated. The deterministic components are estimated from all phones present in the speech corpus, and these are systematically arranged in the form of a decision tree (described in Sect. 5.2.2). The noise components are parameterized in terms of spectral and amplitude envelopes (detailed in Sect. 5.2.4). Harmonic to noise ratio is computed as the ratio of the energy of deterministic and noise components. During synthesis, for every phone, the suitable deterministic component is chosen from the leaf of the decision tree, and the noise component is obtained by imposing the target spectral and amplitude envelopes on the natural instance of the noise signal. The procedure for selecting the natural instance of the noise signal is provided in Sect. 5.2.3. The energy of deterministic component is kept constant throughout the phone, and the energy of noise component is adjusted according to the target HNR. The deterministic and the noise components are pitch-synchronously overlap-added to generate the excitation signal of a phone.

The proposed method is illustrated by considering the excitation signal of a phone /ey/ obtained from SLT speaker. Figure 5.10a, b show the natural speech and the excitation signal of a phone /ey/. The excitation signal consists of 18 pitch-synchronous residual frames. The excitation signal is modeled by using the proposed time-domain deterministic plus noise model. HNR, energy, and spectral and amplitude envelope parameters of noise components are estimated. The estimated deterministic component and a single instance of noise component are shown in Fig. 5.11. The excitation signal is reconstructed from the estimated parameters. The reconstructed excitation signal and the corresponding synthesized speech signal are shown in Fig. 5.10d and c, respectively. To further analyze

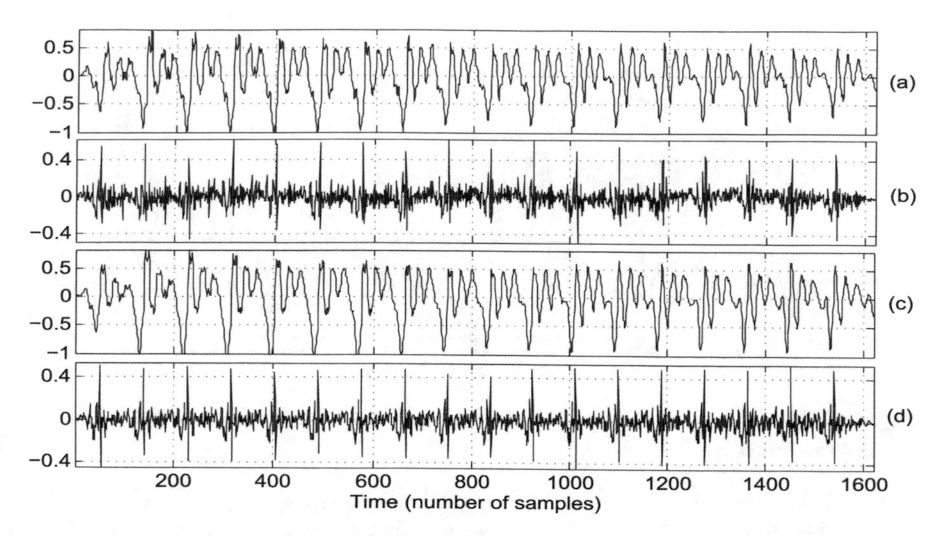

Fig. 5.10 (**a**) The original speech signal and (**b**) the corresponding excitation signal of a phone /ey/ obtained from SLT speaker. (**c**) The synthesized speech signal and (**d**) the reconstructed excitation signal using the deterministic and noise components

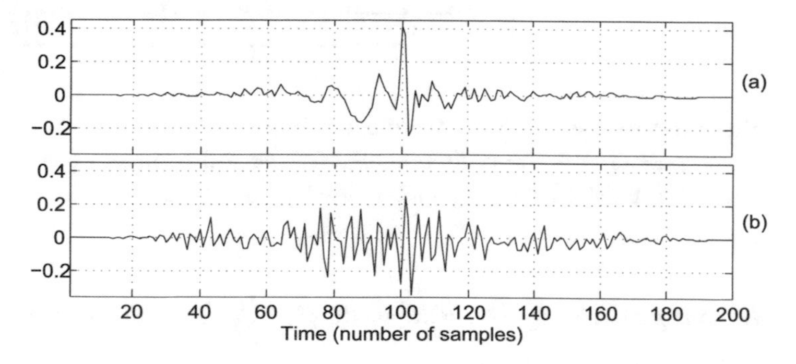

Fig. 5.11 (**a**) The estimated deterministic component and (**b**) single instance of noise component for the phone /ey/

the effect of the number of residual frames on the estimation of deterministic component, the different numbers of residual frames (5, 10, and 15) are considered for the decomposition of deterministic and noise component. Figure 5.12a–c shows the speech waveforms synthesized using the deterministic component estimated from 5, 10, and 15 residual frames, respectively. The similarity of the synthesized speech waveforms and the natural speech is objectively measured using log-spectral distance (LSD). Log-spectral distances computed between the natural speech and the speech synthesized by considering the deterministic component estimated from the different number of residual frames of a phone are shown in Table 5.2. From the table, it can be observed that the LSD value decreases with increase in the number of residual frames considered for the estimation of deterministic component. With the increase in the number of residual frames, the noise components are canceled

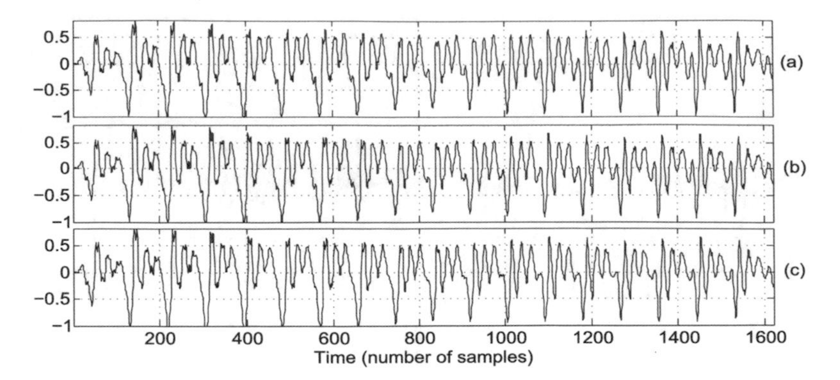

Fig. 5.12 Speech synthesized using the deterministic component estimated from (**a**) 5, (**b**) 10, and (**c**) 15 residual frames

Table 5.2 Log-spectral distances computed by considering the deterministic component estimated from different numbers of residual frames

Number of residual frames considered for computing the deterministic component	LSD
5	0.9470
10	0.9345
15	0.9310
All	0.9301

out, and the estimated deterministic component is precise. The steps involved in modeling and generation of the excitation signal under time-domain deterministic plus noise model framework are explained in the following section.

5.2.1 *Decomposition of Deterministic and Noise Components*

For every speech utterance, the corresponding phonetic transcriptions are available in the database. The phonetic transcriptions provide the sequence of phones present in the speech utterance and their duration boundaries (start and end time stamps). Using the segmentation boundaries, the pitch-synchronous residual frames of a phone are grouped together. The deterministic and noise components are computed using Eqs. (5.4) and (5.5).

5.2.2 *Clustering the Deterministic Components*

During synthesis, the deterministic components of all phones are available for the generation of the excitation signal. The most appropriate deterministic component which best matches with the given target phone specification is selected from the database. To accomplish this, the deterministic components are grouped into a number of clusters. All the units present in the cluster have similar acoustic characteristics. The procedure followed for clustering the deterministic components is same as that of clustering the optimal residual frames (described in Sect. 5.1.2). The set of acoustic features used for describing the optimal residual frame is also used for parameterizing the deterministic components. By considering the deterministic components of all the phones present in the database, a single decision tree is constructed whose nodes have questions related to the features of phones and leaves contain the clusters of deterministic components. The minimum number of units present in the cluster is specified to be 10.

5.2.3 Storing Natural Instance of Noise Signal Along with the Deterministic Component

During synthesis, the deterministic component chosen from the leaf of the decision tree is added to the noise component of desired spectral and amplitude envelopes to generate the excitation signal. Most of the existing methods have used white noise to generate the noise component of excitation [7, 8]. The spectral and amplitude envelopes estimated from the noise components of phone /ey/ (shown in Fig. 5.10) are imposed on the white noise. The resulting synthesized speech and the excitation signal are shown in Fig. 5.13. The speech synthesized from the white noise (Fig. 5.13a termed as SS_{ew}) and the speech synthesized from the natural instance of noise signal (Fig. 5.10c termed as SS_{en}) are compared with the natural speech signal (Fig. 5.10a termed as NS). The log-spectral distance between SS_{ew} and NS is observed to be 0.9443. This is higher than the LSD between SS_{en} and NS (given in the last row of Table 5.2). This indicates that the SS_{ew} is deviating more from the NS compared to the SS_{en}. From the informal listening tests of many instances of SS_{ew} and SS_{en} of phone /ey/, it is observed that the SS_{en} is perceptually more close to the NS. The similar results are observed for different phonetic classes.

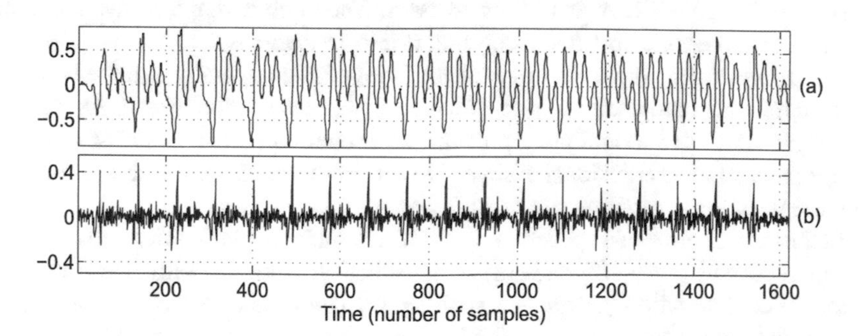

Fig. 5.13 (**a**) The synthesized speech signal and (**b**) the excitation signal obtained by using white noise for generating the noise component

Maximum fusion is achieved between the deterministic and noise components when the real instance of the noise signal is used. Hence, in addition to every deterministic component, a single instance of a noise signal is stored in the decision tree. During synthesis, the target spectral and amplitude envelopes are imposed on this noise signal to generate the desired noise component.

The single instance of the noise signal that is stored along with the deterministic component is chosen from the noise components estimated from the excitation signal of a phone. For example, if there are 20 pitch-synchronous residual frames in a phone, on deterministic plus noise decomposition, 1 deterministic and 20 noise components are obtained. Out of 20 noise components, one appropriate noise component is chosen. From every noise component of a phone, the spectral and amplitude envelopes are computed. By considering the spectral and the amplitude envelopes of all noise components, average spectral and amplitude envelopes are calculated. The noise component having the spectral and amplitude envelopes close to the average spectral and amplitude envelopes is considered as the appropriate noise component. The noise component is energy normalized and stored along with the deterministic component. This noise component is termed as the natural instance of the noise signal. Similarly for every deterministic component estimated from all phones present in the speech corpus, appropriate noise signals are stored. The selection of noise component that is close to average spectral and amplitude envelopes ensures minimum modification of these envelopes during synthesis.

5.2.4 *Parameterizing the Noise Components*

The noise component is parameterized in terms of its spectral and amplitude envelopes. The steps followed for the parameterization of the noise component are similar to Sect. 4.2.4. The spectral envelope of the noise component is represented using 10th order LPC coefficients. The LPC coefficients are converted to LSF coefficients. The amplitude envelope $(a(n))$ of the noise component is smoothed by filtering the absolute value of the noise component $(u(n))$ with a moving average filter of order $2N + 1$. N is chosen to be 8. The overall shape of the amplitude envelope is represented by downsampling it into 15 samples.

The spectral and amplitude envelopes of the noise component and the harmonic to noise ratio are computed for every pitch-synchronous residual frame. As followed in parametric source modeling approach, the excitation parameters extracted from the pitch-synchronous residual frames present in every 25 ms frame are averaged and assigned as the parameters of that frame.

In the case of unvoiced speech, only MGC coefficients (34th order with $\alpha = 0.42$, $Fs = 16$ kHz, and $\gamma = -1/3$) and energy of excitation signal are extracted from every frame. Other excitation parameters such as F_0, HNR, and spectral and amplitude envelopes of noise are set to zero. We have also examined the excitation parameters of unvoiced speech with their mean values instead of zero. But noticeable difference in the synthesized speech was not observed while using the excitation parameters as

Table 5.3 Source features and the number of parameters

Features	Parameters per frame
Pitch	1
Energy	1
HNR	1
Noise spectrum	10
Noise amplitude envelope	15

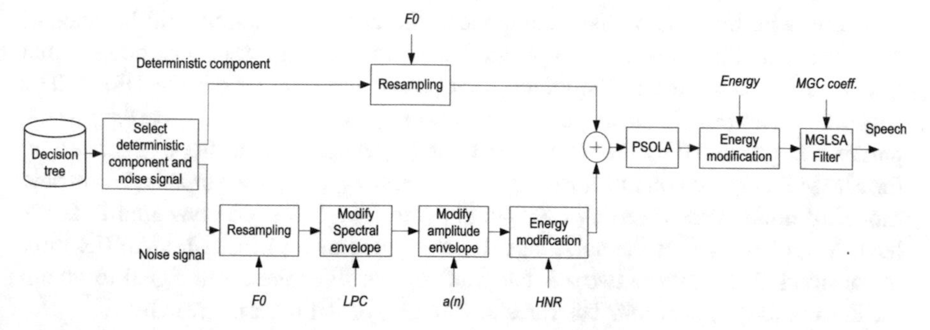

Fig. 5.14 Block diagram showing different synthesis stages in the proposed time-domain deterministic plus noise model-based hybrid source model. The parameters generated by the HMMs are shown in italics

zero or mean values during unvoiced regions. The source features considered in this work are given in Table 5.3. The source features are modeled under the framework of HMM.

5.2.5 Speech Synthesis Using the Proposed Time-Domain Deterministic Plus Noise Model-Based Hybrid Source Model

The block diagram consisting of different synthesis stages is shown in Fig. 5.14. The excitation signal is generated separately for voiced and unvoiced phones. For the voiced phones, the excitation signal is generated in two stages: (1) selection of the suitable deterministic component from the leaf of the decision tree and (2) generation of the noise component. For unvoiced phone, white noise is used as the excitation signal.

During synthesis, for the given input voiced phone, the most appropriate deterministic component is chosen from the leaf of the decision tree. Based on the positional and contextual features of a phone, a cluster of units present at the leaf of the decision tree is selected by answering the questions at each node. To select the most appropriate unit from the cluster of candidate units, target and concatenation costs are computed. The target and concatenation costs are calculated as the weighted sum of subcosts. The features used as the subcosts of target and concatenation costs are same as that used for selecting the optimal residual frames (given in Sect. 5.1.3). The weights of the target and concatenation subcosts are determined manually, based on informal listening tests. The sequence of candidate units which has the lowest sum of target and concatenation costs is chosen as the best sequence of deterministic components of the utterance. The chosen deterministic components are resampled according to the target pitch contour generated by the HMM.

Along with the deterministic component, the corresponding natural instance of the noise signal is also chosen from the leaf of the decision tree. The noise signal is resampled according to the target pitch period generated from the HMM. The spectral and amplitude envelopes of the noise signal are modified according to the parameters generated by the HMM. First, the target spectral envelope generated by the HMM is imposed on the noise signal. The target spectral envelope is the all-pole model of noise represented by LSF coefficients. The LSFs are converted to LPCs (a_k). All-pole model of the noise signal is evaluated using LPCs (b_k). An IIR filter is constructed from these two all-pole models that filter the noise signal to obtain the desired target spectrum. The transfer function of IIR filter is given by

$$H(z) = \frac{(1 - O(z))}{(1 - G(z))} \tag{5.6}$$

where $O(z) = \sum_{k=1}^{p} b_k z^{-k}$ and $G(z) = \sum_{k=1}^{p} a_k z^{-k}$ are the FIR filters obtained from the LPCs of the noise signal and target spectral envelope, respectively. The IIR filtering operation in Eq. (5.6) can be represented as a cascade of two filtering operations. The first operation is FIR filtering operation that is equivalent to LPC analysis. This operation results in the flattening of the spectrum of the noise signal. The next operation is IIR filtering that is equivalent to LPC synthesis. In this operation, the desired target spectrum is applied on the spectrally flat noise signal.

The target amplitude envelope ($a(n)$) generated by the HMM is imposed on the IIR filtered noise signal. The target amplitude envelope that is represented by 15 samples is upsampled to the required target pitch period. The amplitude envelope of the IIR filtered noise signal is also computed. The target envelope is imposed on the IIR filtered noise signal by compensating the difference between two envelopes. The spectrum and amplitude envelopes modified noise signal is the required target noise component. The energy of the target noise component is modified according to the generated HNR, keeping the energy of the deterministic component constant throughout the phone. Both deterministic and noise components are added together. The combination of deterministic and noise components are pitch-synchronously overlap-added to construct the excitation signal. The gain of the excitation signal is matched according to the energy measure generated by the HMM. The resulting signal is used as the excitation of voiced frames. For unvoiced frames, the energy of white noise is modified according to the generated energy measure. The generated excitation is finally given as input to MGLSA filter to produce the speech.

5.2.6 Evaluation

As followed in previous approaches, the proposed method is evaluated using four English speakers (SLT, CLB, AWB, and KSP) from CMU Arctic speech database [5]. Subjective evaluation is performed using two measures, namely, CMOS and preference tests. Before comparing the proposed method with the existing source modeling methods, the proposed method is compared with its three variants, namely, (1) source model using single deterministic component for all phones (single-det), (2) source model using one deterministic component for every phone (multiple-det), and (3) source model using pruned-clusters to generate the excitation signal. The pruned-clusters contain only one deterministic component which is close to cluster center at every leaf of the decision tree. In the proposed method, all units present at the leaves of the decision tree are utilized for the generation of the excitation signal (also referred as full-cluster).

CMOS with 95% confidence intervals and preference scores obtained for four speakers are shown in Figs. 5.15 and 5.16, respectively. In figures, Set1, Set2, and Set3 indicate the comparison of the proposed method with the single-det, multiple-det, and pruned-cluster versions, respectively. From figures, it can be observed that the proposed method is better than the single-det for all speakers. On comparison of the proposed method with the multiple-det, it can be observed that the CMOS and preference scores are less than the single-det. This implies that by utilizing multiple deterministic components, the quality of the synthesized speech has improved compared to using single deterministic component. Regarding the comparison of the proposed method with the pruned-cluster, it can be observed that the CMOS and

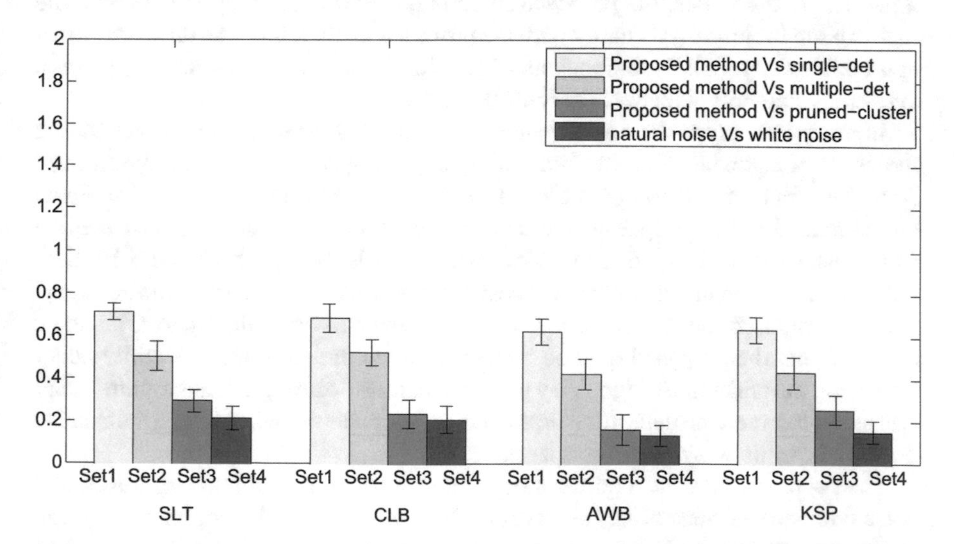

Fig. 5.15 CMOS scores with 95% confidence intervals obtained by comparing the proposed method with its four variants

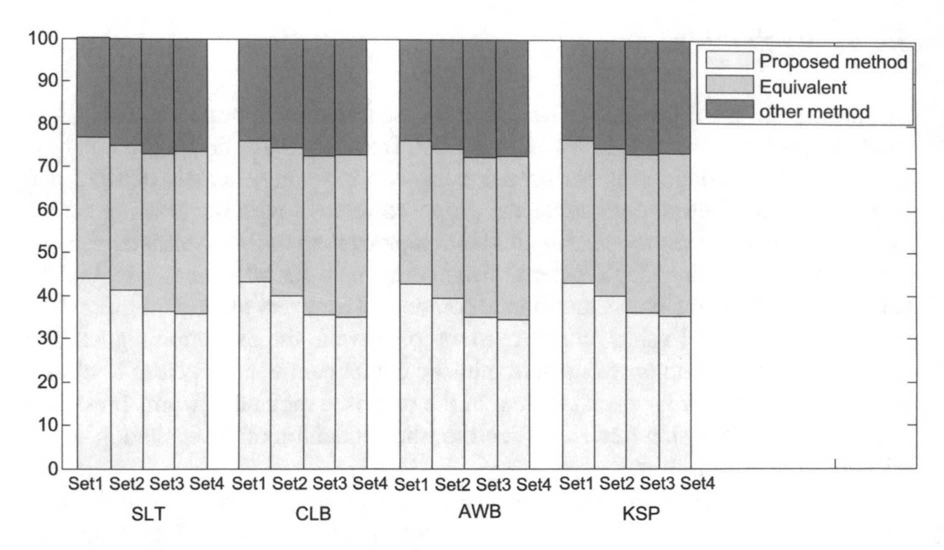

Fig. 5.16 Preference scores obtained by comparing the proposed method with its four variants

preference scores are less than that of single-det and multiple-det. This shows that the quality of the speech synthesized by the pruned-cluster is better than single-det and multiple-det. The pruned-cluster provides multiple deterministic components with different contexts for the generation of the excitation signal. During synthesis, an appropriate deterministic component is selected from the leaf of the decision tree by answering all questions related to the positional and contextual features of a phone. On the whole, the proposed method is better than its three variants. The main reason for improved quality with the proposed method is due to the selection of appropriate deterministic components which closely matches the target specification and varies smoothly with the consecutive phones.

In order to analyze the effectiveness of the natural noise signal for generating the noise component of the excitation signal, the speech utterances synthesized from the proposed method (full-cluster version) by using the natural noise signal are compared with the speech utterances synthesized from the proposed method (full-cluster version) by using the white noise signals. In Figs. 5.15 and 5.16, Set4 indicates the comparison of the proposed method using the natural noise signal with the proposed method using the white noise signal. From the figures, it can be observed that the proposed method using the natural noise signal is slightly better than using the white noise signal. By using the natural noise signal, maximum fusion is ensured between the deterministic and noise components and, hence, results in the improved quality of synthesized speech.

The proposed method requires more memory for storing the deterministic and noise components present at the leaves of the decision tree. The memory required by the proposed method and its three variants is analyzed. For SLT speaker, each of the deterministic and noise component waveforms is represented using 200 samples (normalized pitch period). If each sample is represented in floating point format that

Table 5.4 Number of units and memory required to store the pruned-cluster and full-cluster versions of the proposed method

Speaker	Number of units in full-cluster	Number of units in pruned-cluster	Memory of full-cluster (MB)	Memory of pruned-cluster (MB)
SLT	24,451	1528	39.12	1.95
CLB	26,122	1694	41.79	2.08
AWB	22,528	1370	36.04	1.80
KSP	21,746	1195	34.79	1.73

requires 4 bytes, then the memory needed to store one pair of deterministic and noise component waveforms is $400 \times 4 = 1600$ bytes. In single-det version, only one pair of deterministic and noise component waveforms is used, and it requires 1.6 kilobyte (KB) of memory. As the total number of unique voiced phones present in the database considered is 28, the memory required to store multiple-det version is $28 \times 1.6\,\text{KB} = 44.8\,\text{KB}$. For the pruned-cluster and full-cluster versions of the proposed method, the number of units and the memory required to store the pruned-cluster and full-cluster versions of the proposed method are given in Table 5.4. From the table, it can be observed that substantial decrease (about 95%) in the memory can be obtained from the pruned-cluster compared to the full-cluster version of the proposed method.

Two versions of the proposed source model, namely, full-cluster and pruned-cluster, are compared with three other existing methods, namely, Pulse-HTS, STRAIGHT-HTS [9], and DSM-HTS [8]. The details of each of these source modeling methods are provided in Sect. 4.2.6. In the remaining part of the section, two versions of the proposed method (i.e., full-cluster and pruned-cluster) are compared with each of the three existing systems separately.

The speech synthesized by the proposed method is compared with the Pulse-HTS through CMOS and preference tests. The CMOS with 95% confidence intervals and preference scores are provided in Figs. 5.17 and 5.18, respectively. In the figures, Set1 indicates the comparison of the full-cluster version of the proposed method with the Pulse-HTS, and Set2 indicates the comparison of the pruned-cluster version of the proposed method with the Pulse-HTS. From Fig. 5.17, it can be observed that the CMOS scores vary between 1.5 and 2 for both male and female speakers. The preference scores provided in Fig. 5.18 show that the subjects preferred the proposed method for about 70% of the cases. This indicates that the quality of the speech synthesized by the proposed method (both full-cluster and pruned-cluster) is clearly superior compared to the Pulse-HTS. The subjects clearly noticed that the speech synthesized from the Pulse-HTS is artificial and unnatural. The excitation signals generated using the combination of deterministic and noise components are much better than the sequence of pulses.

The CMOS with 95% confidence intervals and preference scores obtained by comparing the proposed method with the STRAIGHT-HTS are provided in Figs. 5.19 and 5.20, respectively. In figures, Set1 indicates the comparison of the full-cluster version of the proposed method with the STRAIGHT-HTS, and Set2

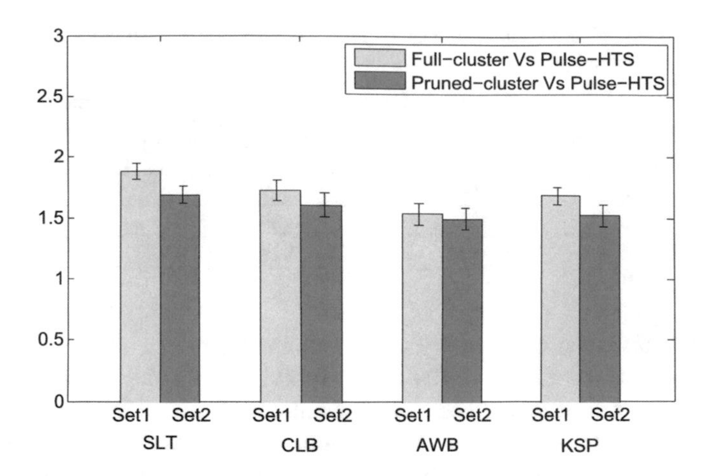

Fig. 5.17 CMOS scores with 95% confidence intervals obtained by comparing the proposed method with the Pulse-HTS

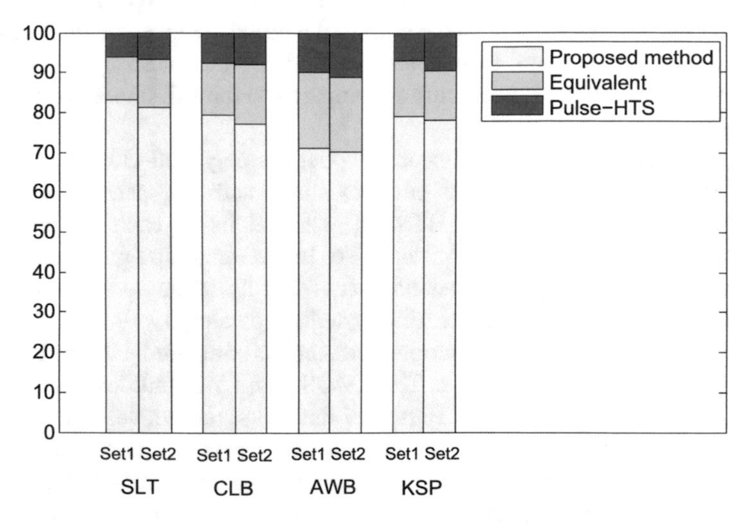

Fig. 5.18 Preference scores obtained by comparing the proposed method with the Pulse-HTS

indicates the comparison of the pruned-cluster version of the proposed method with the STRAIGHT-HTS. From Fig. 5.19, it can be observed that the CMOS scores of both Set1 and Set2 vary between 0.7 and 1 for both male and female speakers. The preference scores provided in Fig. 5.20 show that the subjects preferred the proposed method for about 40% of the cases and preferred the STRAIGHT-HTS for about 25% of the cases. Both measures indicate that the proposed method (both full-cluster and pruned-cluster) is better than the STRAIGHT-HTS. The STRAIGHT vocoder uses the mixed excitation parameters to model and generate the voice source signal. The proposed method uses the real instances of deterministic and noise components for the excitation signal generation. This approach preserves some of the detailed

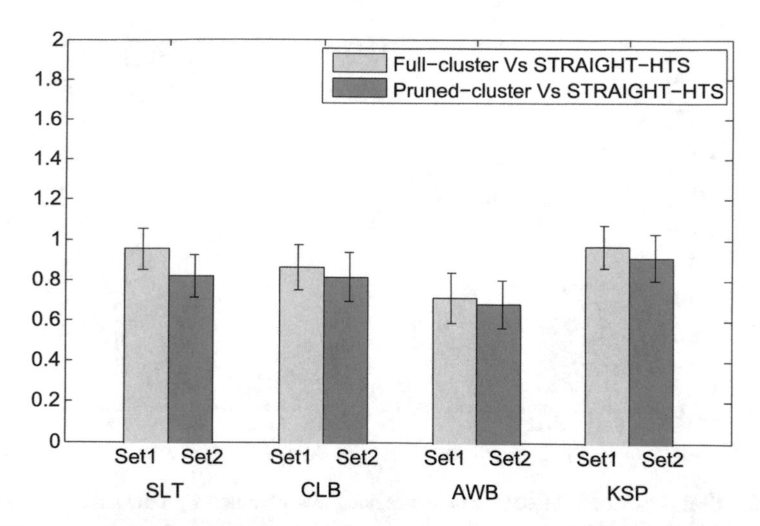

Fig. 5.19 CMOS scores with 95% confidence intervals obtained by comparing the proposed method with the STRAIGHT-HTS

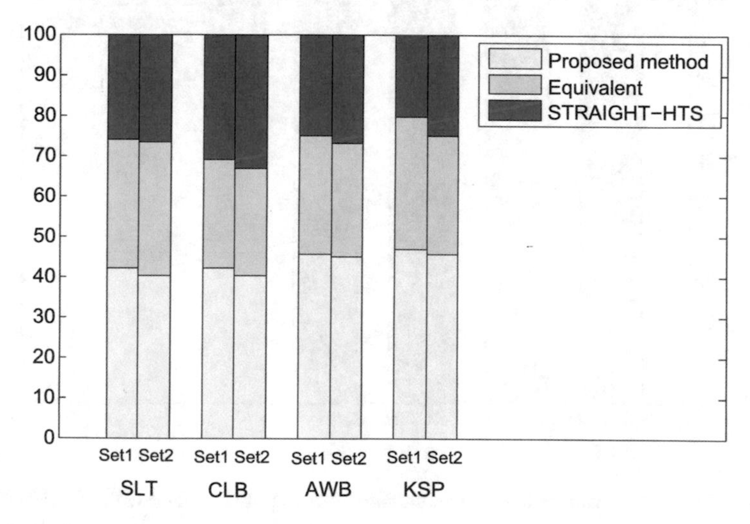

Fig. 5.20 Preference scores obtained by comparing the proposed method with the STRAIGHT-HTS

structure of the natural excitation which cannot be modeled accurately. Hence, the generated excitation signal is close to the natural source signal.

The speech synthesized by the proposed method is compared with the DSM-HTS. The CMOS with 95% confidence intervals and preference scores are provided in Figs. 5.21 and 5.22, respectively. In figures, Set1 indicates the comparison of the full-cluster version of the proposed method with the DSM-HTS, and Set2 indicates the comparison of the pruned-cluster version of the proposed method with the DSM-HTS. From Fig. 5.21, it can be observed that the CMOS scores are varying between

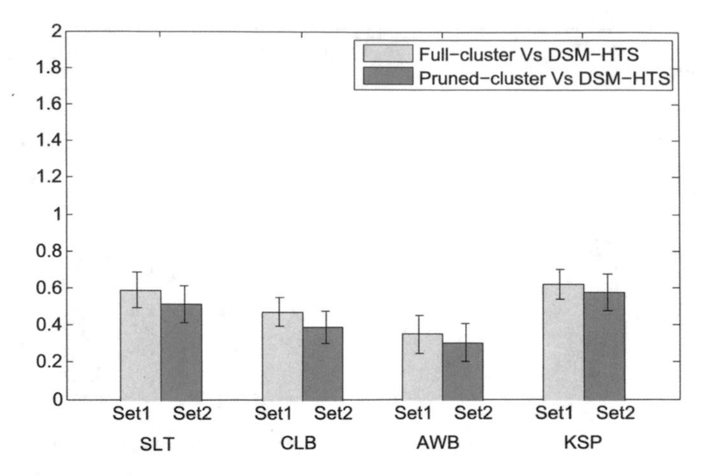

Fig. 5.21 CMOS scores with 95% confidence intervals obtained by comparing the proposed method with the DSM-HTS

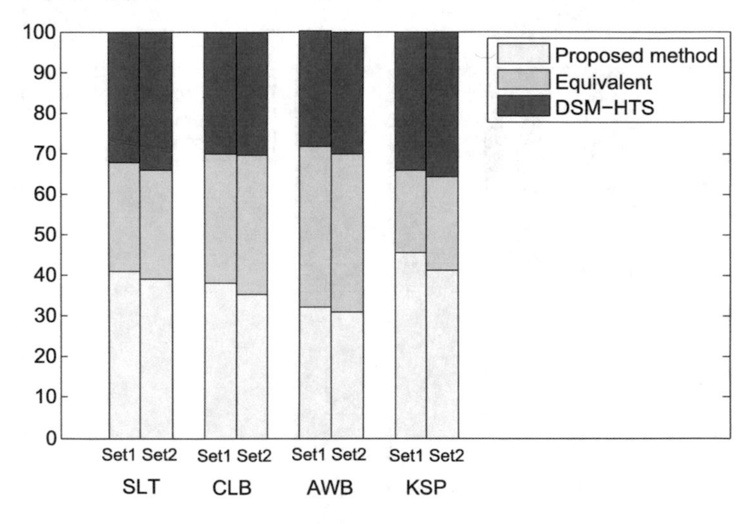

Fig. 5.22 Preference scores obtained by comparing the proposed method with the DSM-HTS

0.3 and 0.6 for both male and female speakers. The preference scores provided in Fig. 5.22 indicate that the subjects preferred the proposed method for about 40% of the cases and preferred the DSM-HTS for about 30% of the cases. Both measures show that the proposed method (both full-cluster and pruned-cluster) is slightly better than the DSM-HTS. The DSM-HTS utilizes the single instance of deterministic component (first eigenvector), and the average spectral and amplitude envelope modulated white Gaussian noise to generate the excitation signal. This kind of modeling cannot incorporate the natural variations of the excitation signal from one pitch-synchronous residual frame to another. In the proposed method, the deterministic component that is constant for the entire duration of the phone is added to the noise component that varies with every cycle of the pitch-synchronous residual frame. This results in the incorporation of characteristics of the real voice source in the excitation signal.

From the CMOS and preference scores obtained by comparing the proposed parametric (based on the analysis of characteristics of residual frames) and hybrid source models with the existing source models, it can be understood that the hybrid method is slightly better than the parametric method. To explicitly perceive the effectiveness of the hybrid method over the parametric method, the two source modeling methods are compared by using CMOS and preference scores. In the CMOS, the positive score indicates that the hybrid method is better than the parametric method, and the negative score indicates the opposite. The CMOS with 95% confidence intervals and the preference scores are provided in Figs. 5.23 and 5.24,

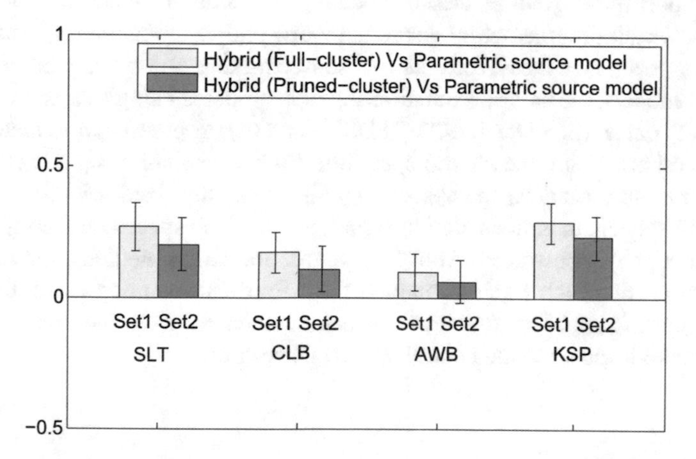

Fig. 5.23 CMOS scores with 95% confidence intervals obtained by comparing the proposed hybrid method with the parametric method

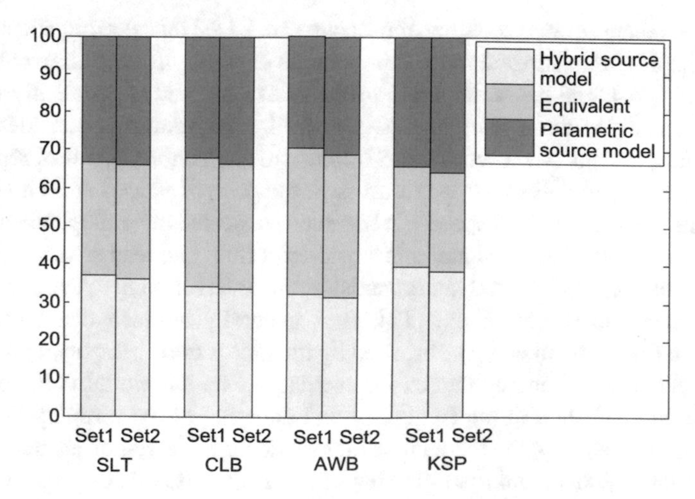

Fig. 5.24 Preference scores obtained by comparing the proposed hybrid method with the parametric method

respectively. In the figures, Set1 indicates the comparison of the full-cluster version of the hybrid method with the parametric method, and Set2 indicates the comparison of the pruned-cluster version of the hybrid method with the parametric method. From Fig. 5.23, it can be observed that the CMOS score is positive and close to zero for all speakers. Compared to the pruned-cluster version, the full-cluster version has higher CMOS and preference scores. From both CMOS and preference scores, it can be observed that the full-cluster version of the hybrid method is slightly better than the parametric method, whereas the pruned-cluster version of the hybrid method is very much close or almost equal to the parametric method. In the pruned-cluster version of the hybrid method, the confidence intervals of the CMOS scores are very close to zero. For AWB speaker, the confidence interval is going toward negative scale. The synthesized speech samples of the proposed hybrid source modeling approach and three existing source modeling methods are made available online at http://www.sit.iitkgp.ernet.in/~ksrao/narendra-phd-demo/narendra_hsm.html.

In order to analyze the effectiveness of source models without any influence from statistical models, the natural excitation signal is modeled using four source models, namely, (1) pulse, (2) STRAIGHT, (3) DSM, and (4) proposed hybrid method (full-cluster version). Using the natural spectrum, F0 and excitation signals constructed from four source models, the speech signals are synthesized. Figure 5.25 shows the natural speech, natural excitation signal, synthesized speech, and corresponding excitation signals constructed from four source modeling methods. On comparing the synthesized speech signals obtained from four source models with the natural speech signal, it is observed that the speech synthesized from the proposed hybrid source model is closer to the natural speech waveform.

5.2.7 *Discussion*

Most of the recent source modeling approaches are based on deterministic plus noise model. The deterministic and noise components capture different characteristics of the source signal, and hence, separate methods can be used to efficiently represent and model each of these components. Generally, the source is considered to be entirely independent of the vocal-tract filter, and the source signal is represented as a sequence of impulse-type waveforms in the case of voiced speech and white noise in the case of unvoiced speech. It has been suspected that independent analysis and modeling of the source signal and vocal-tract filter can lead to the reduction in speech quality [10, 11]. Usually, there exists some interaction between the activities of vocal tract and source signal [12]. It is generally realized that there can be appreciable first formant energy absorbed by the glottis during the open phase of the glottal cycle, and this energy can cause oscillations on the glottal volume velocity waveform and a change in the frequency and damping of the formant. Due to this interaction, the shapes of the residual frames particularly the region around GCI vary slowly within a phone, and also the shapes of residual frames depend on the type of phone and its context. To exploit these phone-dependent characteristics of the

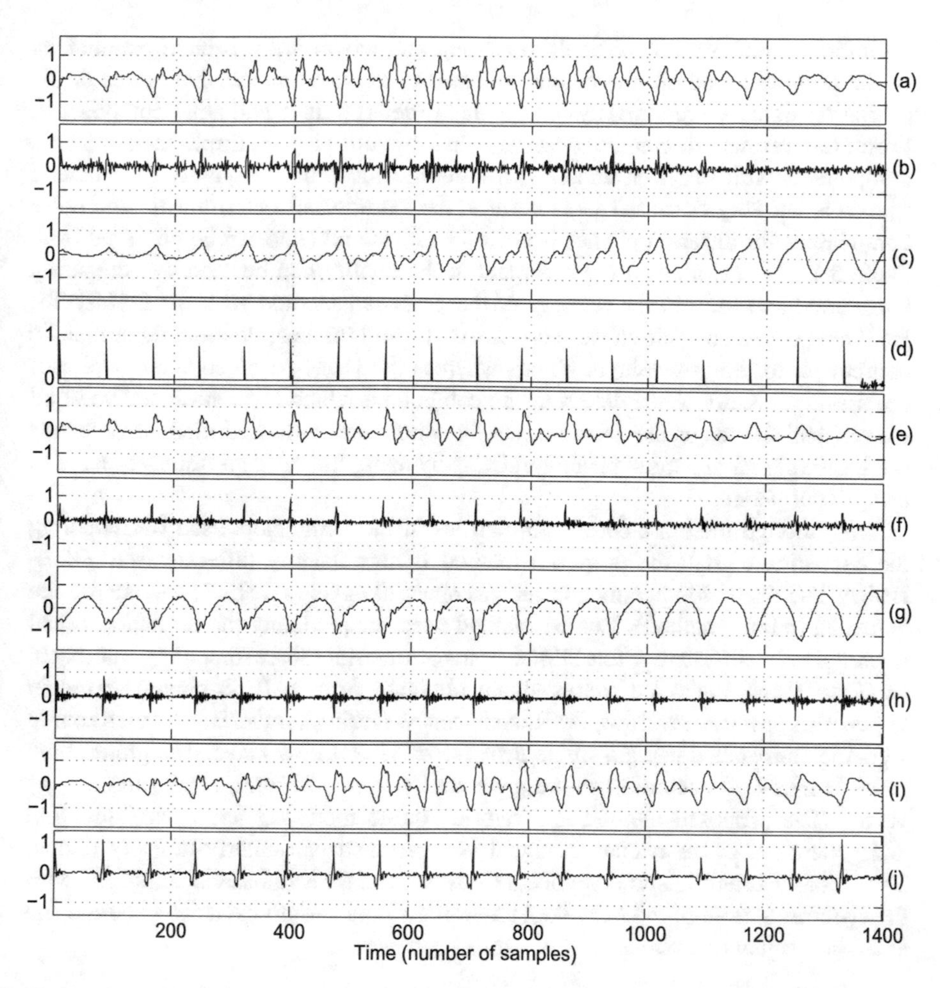

Fig. 5.25 (**a**) Natural speech, (**b**) excitation signal, (**c**) speech synthesized by pulse-based source model, (**d**) excitation signal generated by pulse-based source model, (**e**) speech synthesized by STRAIGHT source model, (**f**) excitation signal generated by STRAIGHT source model, (**g**) speech synthesized by DSM-based source model, (**h**) excitation signal generated by DSM-based source model, (**i**) speech synthesized by proposed source model, and (**j**) excitation signal generated by proposed hybrid source model

residual frame, the deterministic components estimated from all the phones present in the database are clustered in the form of a decision tree. During synthesis, a suitable deterministic component is selected from the decision tree that best matches with the input phone and its context.

The evaluation results demonstrate that the modeling of the source signal using time-domain deterministic plus noise model can improve the quality of synthetic speech. In the proposed method, the utilization of natural residual frames and natural noise signal can preserve some of the finer detailed structure of the natural excitation, which will be difficult to capture and model with discrete parameters. In the proposed method, the natural variations present in successive cycles of the excitation signal are retained by combining the deterministic and noise components. From the results, it can be observed that by utilizing multiple deterministic components with different contexts, the improvement in the quality of synthetic speech is not very drastic. This is because the variation of deterministic and noise components for different phones is not very substantial (this can be observed from Table 5.1). Hence, a small improvement in the quality can be seen by comparing the pruned and full-cluster versions of the proposed method with the DSM-HTS. Both the pruned and full-cluster versions try to find the best possible deterministic component for the given input phone, whereas the DSM-based source model uses the first eigenvector as the deterministic component which is a representative of all the residual frames of the database. On the whole, the proposed method generates the source signal in a more meaningful way by using the residual frames specific to every input phone.

Even though there are certain similarities in the parameters used for modeling the excitation signal, the proposed method is significantly different from Glott-HMM. The excitation parameters such as spectral envelope, HNR, and energy are extracted in both methods. But the method used for modeling the excitation signal is completely different. Glott-HMM is a parametric source modeling approach. In Glott-HMM, the voice source signal obtained from IAIF is parameterized in terms of its spectral envelope, HNR, and energy. During synthesis, the voice source signal is constructed using a single instance of glottal flow pulse. The glottal flow pulse is modified according to the generated source spectrum, HNR, and energy values. The proposed method is a hybrid source modeling approach. Modeling and generation of the excitation signal is performed particularly to every phone. The suitable deterministic component chosen from the database is combined with the spectral and amplitude envelopes modified noise components to generate the excitation signal of a phone.

5.3 Summary

Two hybrid source modeling methods are proposed in this chapter. In the first hybrid source model, the excitation signal of a phone is modeled by using a single optimal residual frame. The optimal residual frames extracted from all phones present in the database are efficiently grouped in the form of a decision tree. During synthesis, the appropriate residual frames are chosen based on target and concatenation costs. The CMOS and preference scores indicate the significance of the proposed source modeling over traditional pulse excitation. To accurately represent each of the residual frames and incorporate cycle to cycle variations in the residual frames of the excitation signal, a hybrid source model is proposed based on time-domain deterministic plus noise model. In this hybrid source model, the residual frames of the excitation signal are modeled as a combination of deterministic and noise components. The deterministic components are efficiently arranged in the form of a decision tree. The noise components are parameterized in terms of its spectrum and amplitude envelopes. During synthesis, the deterministic component is added to the spectrum and amplitude envelope modified natural instance of the noise signal to generate the excitation signal. This novel procedure of the generation of excitation signal enables the production of high-quality synthetic speech. The subjective evaluation results showed that the quality of the proposed approach is considerably better compared to three state-of-the-art source modeling methods.

References

1. HMM-based speech synthesis system (HTS) [Online]. http://hts.sp.nitech.ac.jp/
2. L. Breiman, J. Friedman, C.J. Stone, R.A. Olshen, *Classification and Regression Trees* (Wadsworth & Brooks, Pacific Grove, 1984)
3. T. Raitio, A. Suni, H. Pulakka, M. Vainio, P. Alku, Utilizing glottal source pulse library for generating improved excitation signal for HMM-based speech synthesis, in *Proceedings of International Conference on Acoustics, Speech and Signal Processing, (ICASSP)* (2011), pp. 4564–4567
4. R.A. Clark, K. Richmond, S. King, Multisyn: open-domain unit selection for the Festival speech synthesis system. Speech Commun. **49**, 317–330 (2007)
5. CMU ARCTIC speech synthesis databases [Online]. http://festvox.org/cmu_arctic/
6. G. Seshadri, B. Yegnanarayana, Perceived loudness of speech based on the characteristics of glottal excitation source. J. Acoust. Soc. Am. **4**, 2061–2071 (2009)
7. T. Raitio, A. Suni, J. Yamagishi, H. Pulakka, J. Nurminen, M. Vainio, P. Alku, HMM-based speech synthesis utilizing glottal inverse filtering. IEEE Trans. Audio Speech Lang. Process. **19**(1), 153–165 (2011)
8. T. Drugman, T. Dutoit, The deterministic plus stochastic model of the residual signal and its applications. IEEE Trans. Audio Speech Lang. Process. **20**(3), 968–981 (2012)
9. H. Zen, T. Toda, M. Nakamura, K. Tokuda, Details of Nitech HMM-based speech synthesis system for the Blizzard Challenge 2005. IEICE Trans. Inf. Syst. **E90-D**(1), 325–333 (2007)
10. G. Fant, *Acoustic Theory of Speech Production* (Mouton De Gruyter, Berlin, 1960)
11. J.L. Flanagan, *Source-System Interaction in the Vocal Tract*. Ann. New York Acad. Sci. **155**(1), 9–17 (1968)
12. I.R. Titze, B.H. Story, Acoustic interactions of the voice source with the lower vocal tract. J. Acoust. Soc. Am. **101**(4), 2234–2243 (1997)

Chapter 6
Generation of Creaky Voice

6.1 HMM-Based Speech Synthesis System for Generating Modal and Creaky Voices

The block diagram of the HMM-based speech synthesis system including additional modules to generate the creaky voice is shown in Fig. 6.1. A speech database that has a significant amount of creaky voice is considered for developing HTS. From the speech database, the creaky regions are identified using the proposed creaky voice detection method (described in Sect. 6.2). The proposed creaky voice detection method outputs a creaky probability (P_{crk}) for every frame of speech. For every speech utterance present in the database, the corresponding excitation signal is obtained. Using the creaky probability P_{crk}, the creaky regions are identified in the excitation signal. The excitation signal of the normal voice (includes modal voiced and unvoiced) is modeled using the time-domain deterministic plus noise model-based hybrid source model proposed in the previous chapter. The excitation signal of the creaky voice is modeled using the proposed hybrid source model that is an extension of the basic time-domain deterministic plus noise model (explained in Sect. 6.3). The proposed hybrid source model extracts creaky parameters and stores the real segments of creaky residual frames in the database. F0 estimation and voicing decision are performed on every utterance present in the speech database. In the creaky voice, due to low F0 and highly irregular periodicity, F0 estimation methods either output spurious F0 values or incorrectly determine the region to be unvoiced. In this work, F0 estimation and voicing decision are performed based on the strength of instants of significant excitation proposed in Chap. 3. This method has demonstrated its efficiency by identifying both modal and creaky regions as voiced and extracts accurate F0 values. MGC coefficients (34th order with $\alpha = 0.42$, $Fs = 16$ kHz, and $\gamma = -1/3$) which represent the spectrum of speech are extracted from every utterance. The MGC coefficients, F0 with voicing decision, creaky probability, and creaky parameters are modeled using context-dependent HMMs.

K. S. Rao, N. P. Narendra, *Source Modeling Techniques for Quality Enhancement in Statistical Parametric Speech Synthesis*, SpringerBriefs in Speech Technology, https://doi.org/10.1007/978-3-030-02759-9_6

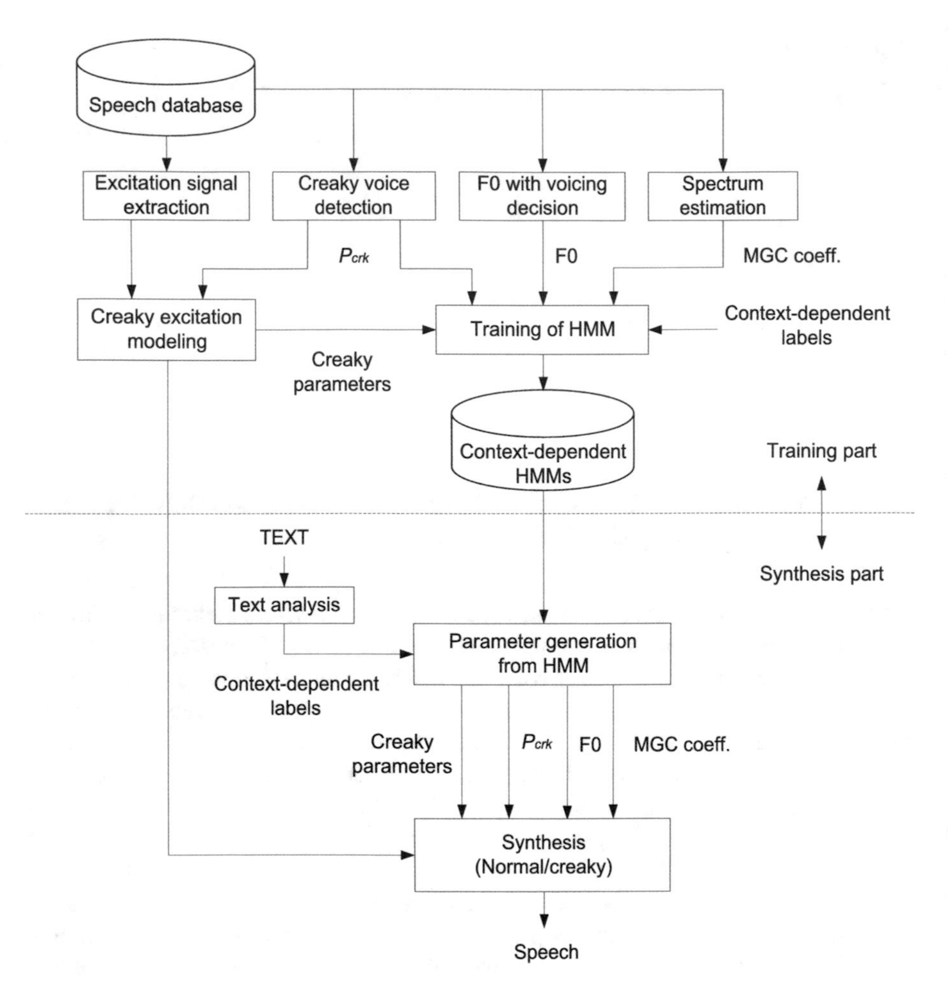

Fig. 6.1 Block diagram of the HMM-based speech synthesis system for generating the modal and creaky voices

During synthesis, input text is converted into a sequence of context-dependent phoneme labels. According to the label sequence, a sentence HMM is constructed by concatenating the context-dependent HMMs. Then, a sequence of parameters is generated from the sentence HMM. Depending on the generated creaky probability, the excitation signal is constructed separately for modal and creaky regions. The excitation signal of the creaky regions is constructed from the generated creaky parameters and the real segments of the residual frames stored in the database. Finally, the speech waveform is synthesized using the generated MGC coefficients and the excitation signal. In the following sections, different issues involved in the synthesis of the creaky voice are described.

6.2 Automatic Detection of Creaky Voice

To generate the creaky voice at appropriate places in the synthesized speech, it is essential to have the information regarding the occurrence of creaky regions during synthesis. To predict the occurrence of creaky regions, the creaky voice needs to be automatically detected from the speech utterances during the training phase. In this work, initially, the source of excitation represented in terms of a set of epoch parameters is analyzed in modal and creaky regions. The epoch parameters are extracted directly from speech signal using zero frequency filtering (ZFF) method with different window sizes. Using the variance of epoch parameters, a neural network classifier is trained to detect the creaky regions. In this section, the description of the proposed creaky voice detection method based on epoch parameters is provided.

Figure 6.2 provides the block diagram of the proposed creaky voice detection method. First, the epoch parameters that include the number of epochs in a frame, the strength of excitation of epochs, and the epoch interval between successive epochs are extracted from the speech signal by using zero frequency filtering method [1]. The complete description of the ZFF method is provided in Chap. 3. The ZFF method consists of two steps. In the first step, the speech signal is passed through a cascade of two zero frequency resonators. In the second step, the output of the cascade of two resonators is subtracted from its local mean to obtain zero frequency filtered signal. The time instants of negative to positive zero crossings of the ZFF signal are called as the epochs. The strength of excitation is computed as the slope of the ZFF signal at each epoch location [2]. The strength of excitation indicates the rate of closure of the vocal folds in each glottal cycle [3]. Sharper closure of the vocal folds corresponds to stronger excitation to the vocal-tract system. The epoch interval is computed as the time interval between successive epochs. The epoch parameters characterizing the source of excitation are analyzed in modal and creaky regions (described in Sect. 6.2.1). Based on the analysis, the variance of epoch parameters is computed (described in Sect. 6.2.2). Using the variance of epoch parameters as input features, a neural network classifier is trained to detect the creaky regions (explained in Sect. 6.2.3).

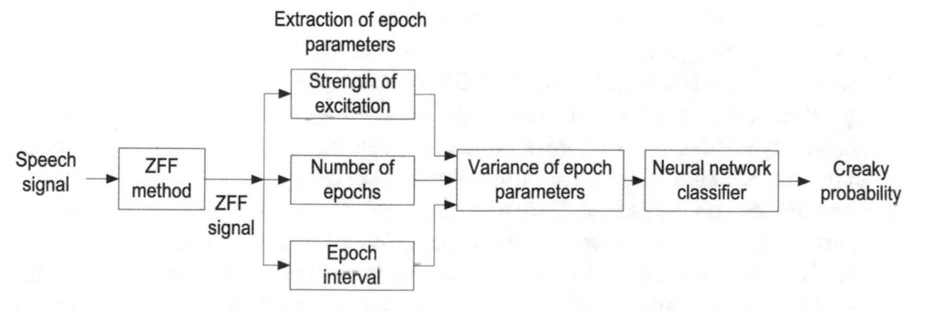

Fig. 6.2 Block diagram of the proposed creaky voice detection method

6.2.1 Analysis of Epoch Parameters

The creaky voice detection method is proposed based on the analysis of the variation of epoch parameters, namely, the number of epochs, the strength of excitation of epochs, and the epoch intervals for different voicing regions. The epoch parameters vary with the size of the window used for local mean subtraction in the ZFF method. In Chap. 3, the variation of the strength of excitation for different window sizes is analyzed for detecting voiced and unvoiced speech regions. In this approach, the variation of epoch parameters for different window sizes is systematically examined for creaky voice detection.

Figure 6.3 presents the strength of excitation (a–d), epoch interval (f–i), and number of epochs (j–m) computed from the speech signal with window sizes of 8, 10, 12, and 14 ms. The speech signal shown in Fig. 6.3e contains three types of regions, namely, unvoiced, modal, and creaky voiced regions. In the unvoiced regions, the vocal folds do not vibrate, and there is no impulse-like excitation. As a result, in the unvoiced regions, the epochs are located at random instants, and the strength of excitation is very low for different window sizes. For the modal regions, the vocal folds vibrate with constant rate, and the most significant impulse-like excitation occurs during GCI. Hence, in the modal regions, the epochs are located at regular intervals, and the strength of excitation is high for different window sizes. In the creaky regions, the vocal folds vibration is low and irregular, and the impulse-like excitation to the vocal-tract system occurs at two instants in each glottal cycle. One impulse-like excitation occurs at the glottal closing instant (primary excitation) and the other at the glottal opening instant (secondary excitation), following a long glottal closed phase. Hence, the epochs are located at uneven locations, and the strength of excitation is higher than the unvoiced regions but lower than the modal regions. On careful analysis of the epoch parameters for modal and creaky regions with different window sizes, three main observations can be drawn.

1. **Strength of excitation (Fig. 6.3a–d):** In the creaky regions, in addition to GCI, the secondary excitation is also present within a glottal cycle. The strength of excitation at GCI is relatively high compared to secondary excitation. As a result for different window sizes, the strength of excitation in successive epochs is varying abruptly. In the modal regions, slow variation in the strength of excitation can be observed across different window sizes.
2. **Epoch interval (Fig. 6.3f–i):** In the modal regions, successive epoch intervals are almost equal or vary slowly. For different window sizes, the variation in the epoch intervals is not significant. In the creaky regions, due to the presence of secondary excitation, successive epoch intervals are unequal. Hence, the epoch intervals vary significantly for different window sizes.
3. **Number of epochs (Fig. 6.3j–m):** In the creaky regions, the secondary excitations having very low strength are detected for lower window sizes and missed for higher window sizes. As a result, the number of epochs in a frame varies for different window sizes. In the modal regions, the secondary excitations are not present. Hence, for different window sizes, the number of epochs in the modal voiced regions does not vary significantly.

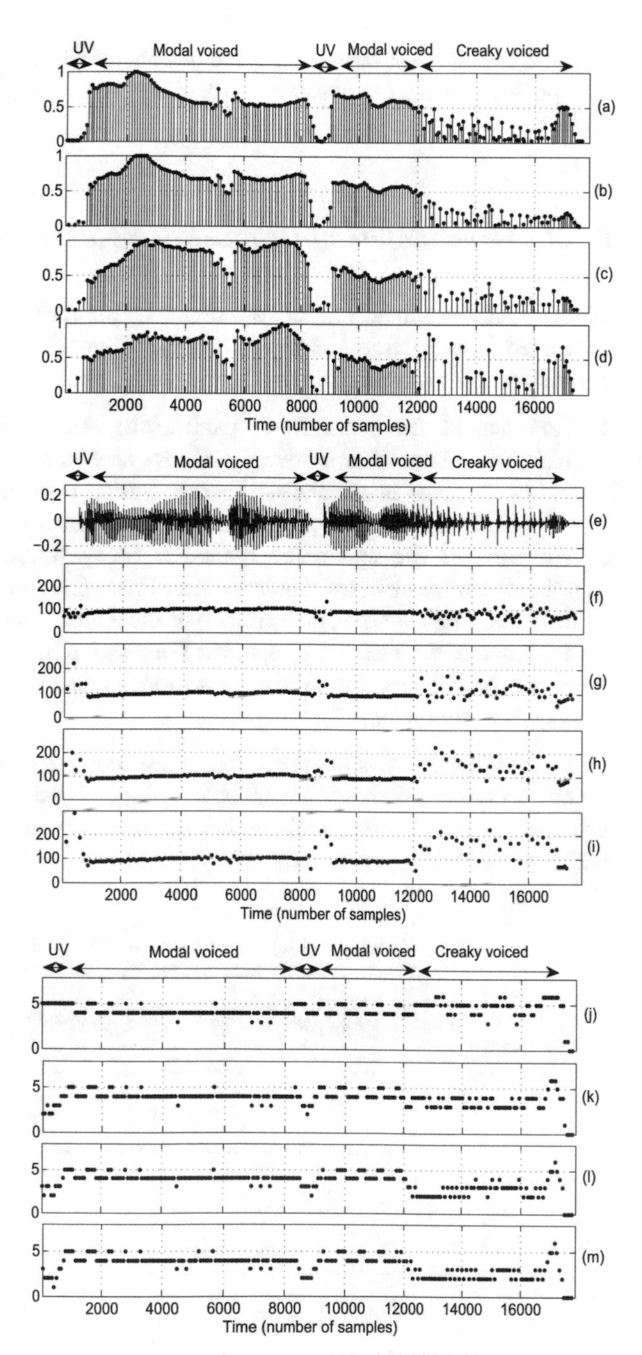

Fig. 6.3 Illustration of the variation of epoch parameters in different voicing regions with different window sizes. Strength of excitation (**a**–**d**), epoch interval (**f**–**i**), and number of epochs (**j**–**m**) computed from the speech signal (**e**) with window sizes of 8, 10, 12, and 14 ms

From the above observations, it can be concluded that for different window sizes, the epoch parameters vary significantly in the creaky regions compared to the modal regions.

6.2.2 Computation of Variance of Epoch Parameters

To differentiate modal and creaky regions, the variance of epoch parameters is computed for every frame of speech. The procedure for finding the variance of epoch parameters is as follows:

1. **Variance of the strength of excitation:** For every frame, the strength of excitation obtained from every window size is normalized between 0 and 1, and its variance is calculated. Average variance of the strength of excitation is computed from the variances obtained for different window sizes.
2. **Variance of the epoch interval:** For every frame, the epoch intervals are collected from different window sizes. From the epoch intervals obtained from different window sizes, the variance of the epoch interval is computed.
3. **Variance of the number of epochs:** For every frame, by considering the number of epochs obtained from different window sizes, the variance of the number of epochs is computed.

All three variances determined from the speech utterance are normalized between 0 and 1. Figure 6.4 shows the speech signal and the variances of the strength of excitation, epoch interval, and number of epochs computed with the window sizes varying from 8 to 14 ms in steps of 1 ms. The procedure for choosing the optimum

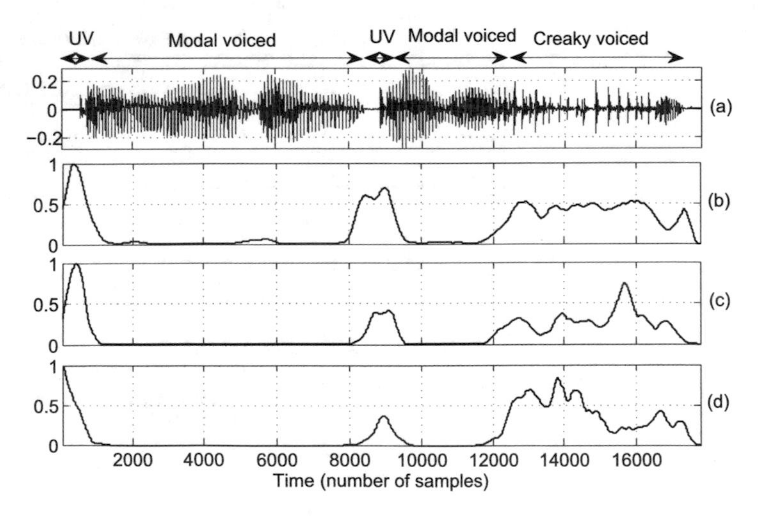

Fig. 6.4 (**a**) Speech signal. Variances of (**b**) strength of excitation, (**c**) epoch interval, and (**d**) number of epochs computed with the window sizes varying from 8 to 14 ms in steps of 1 ms

range of window size is detailed in Sect. 6.2.4. From the figure, it can be observed that the variances have high values in the creaky and unvoiced regions and zero or very low values in the modal regions. Using the voicing detection method at the first stage, the unvoiced regions are removed, and the epoch parameters are extracted only in the voiced regions consisting of modal and creaky regions. For the voicing detection, a method based on the strength of instants of significant excitation proposed in Chap. 3 is used. This method can robustly detect both modal and creaky regions as voiced.

6.2.3 Classification Using Variance of Epoch Parameters

Creaky/non-creaky classification can be performed by applying a threshold on the variance of epoch parameters. Instead of using threshold method, neural network classifier is used. The classifier is configured as a feedforward network consisting of a single hidden layer. All neurons (fixed to 16 in this work) present in the hidden layer utilize a *tanh* transfer function. The output layer consists of a single neuron with a logarithmic sigmoid function suited for a binary decision. The training is performed using a standard error backpropagation algorithm [4]. In this study, the output of the neural network classifier is approximated as a posterior probability or creaky probability P_{crk}.

6.2.4 Performance Evaluation

To evaluate the performance of the proposed method, different speech databases were considered. First, four databases including an American English male speaker (BDL [5]), an Indian English male speaker (KSP [5]), a Finnish female speaker (HS [6]), and a Finnish male speaker (MV [7]) are considered. All four databases were developed to build text-to-speech synthesis systems. A creaky database was developed in two Indian languages, namely, Hindi and Bengali. In each language, single female and male speakers (native voice talents) were used for recording the speech corpora. For both Hindi and Bengali, the text corpus consists of 100 sentences obtained from children stories. The speakers were asked to utter the sentences in news reading style and were asked to intentionally produce the creaky regions at the end of the utterances. One hundred speech utterances obtained from BDL, KSP, HS, MV, Hindi (HF and HM), and Bengali (BF and BM) speech databases were used in the evaluation. All speech utterances were downsampled to a sampling frequency of 16 kHz. To evaluate the detection performance, a reference indicating creaky and non-creaky regions in the speech database was required. Annotation of the creaky regions was performed manually by following a similar approach as described in [8]. The manual annotation of the creaky regions was performed based on the auditory criterion “a rough quality with the additional

Table 6.1 Summary of speech data used for the evaluation of the creaky detection method

Speaker ID	Gender	Language	Creak (%)	Creak duration (s)
BDL	Male	US English	7.6	19.40
KSP	Male	Indian English	2.7	8.34
HS	Female	Finnish	5.8	36.86
MV	Male	Finnish	7.1	31.56
HF	Female	Hindi	9.8	24.81
HM	Male	Hindi	11.1	26.42
BF	Female	Bengali	10.1	20.78
BM	Male	Bengali	12.8	22.18

sensation of repeating impulses." Also, an inspection of waveforms, spectrograms, and F_0 contours was also performed to ensure correct annotation. Table 6.1 provides the summary of speech data used for the evaluation of the creaky detection method.

To assess the performance of the proposed method, three standard frame level metrics are used, namely, true positive rate (TPR, also called recall), false positive rate (FPR), and F1 score. TPR is the proportion of actual creaky frames that are correctly identified. FPR is the percentage of actual non-creaky frames that are wrongly detected as creaky. F1 score is a single metric (bound between 0 and 1) computed using true positives, false positives, and false negatives. The TPR, FPR, and F1 score are obtained from the following equations:

$$\text{TPR} = \frac{\text{True positives}}{\text{True positives} + \text{False negatives}} \tag{6.1}$$

$$\text{FPR} = \frac{\text{False positives}}{\text{False positives} + \text{True negatives}} \tag{6.2}$$

$$\text{F1} = \frac{2 \times \text{True positives}}{2 \times \text{True positives} + \text{False positives} + \text{False negatives}} \tag{6.3}$$

True positives are the number of creaky frames which are correctly detected as creaky. False negatives are the number of creaky frames which are wrongly detected as non-creaky. False positives are the number of non-creaky frames which are wrongly detected as creaky. True negatives are the number of non-creaky frames which are correctly detected as non-creaky. If the creaky voice detection method is better, then TPR and F1 score are higher, and FPR is lower.

In the neural network classifier, for a given input example, the output is the creaky probability, P_{crk}. The standard binary decision is class 1 (i.e., creaky) if $P_{crk} > \alpha$ (otherwise class 0) and α is typically set to 0.5. For skewed data sets (e.g., creak or laughter) which consist of sparse occurrence of a given class to be detected, this setting may not be optimal. Hence, α is varied in the range [0, 1] and set to the value that maximizes the F1 score on the training set. The threshold setting had very low inter-database sensitivity, as all speakers had their best F1 score for α in the vicinity of 0.3. This kind of optimal threshold setting was followed in [9, 10]. Post processing is carried out on the binary decision (creaky or non-creaky) of the classifier. The detection of the creaky regions in very short regions is removed, and nearby adjacent creaky regions are merged by performing a 5-point median filtering to the binary decision. For evaluation purpose, the epoch parameters are extracted for a frame length of 32 ms (assuming lowest F_0 value of creak is 62.5 Hz) and a frame shift of 10 ms.

Optimum Range of Window Size In the Fig. 6.4, the epoch parameters are computed by varying the window size from 8 to 14 ms in steps of 1 ms. With this range of window size, a high distinction of epoch parameters is observed for modal and creaky regions. For different speakers, the optimum range of window size needs to be computed, which results in a higher F1 score and hence better creaky detection. First, by varying the window size from 1 to 20 ms in steps of 1 ms, epoch parameters are computed. With one window size as the center, the epoch parameters of center window size and the epoch parameters of two window sizes before and after the center window size (similar to 5-gram model) are considered. For example, if the center window size is 8 ms, then the epoch parameters of 8, 6, 7, 9, and 10 ms are considered. From the epoch parameters, the variances are computed. Using the variance of epoch parameters, the neural network classifier is trained, and the F1 score is computed. Similarly, with every window size as the center, the variance of epoch parameters are computed, and subsequently, F1 scores are determined. F1 scores obtained with different window sizes as the center for speakers BDL and HS are shown in Fig. 6.5. Average pitch periods of speakers BDL and HS are 5.91 and 5.11 ms, respectively. From the figure, it can be observed that for a window size

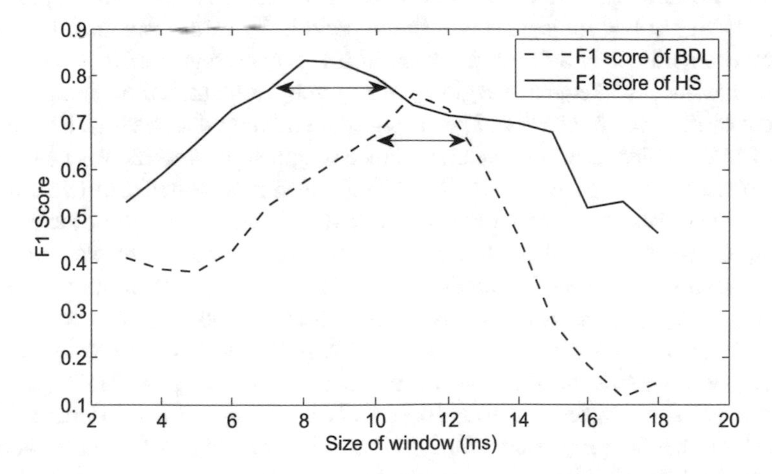

Fig. 6.5 F1 scores obtained with different window sizes as the center for speakers BDL and HS. The double arrow indicates the approximate range of 1.5–2 times pitch period of the speaker, where superior performance is observed

range of approximately 1.5–2 times pitch period of the speaker, F1 score is having high values for both BDL and HS speakers. The similar kind of F1 score curves is obtained for other speakers also. Hence in this work, for computing the variance of epoch parameters, the window size is varied from 1.5 to 2 times pitch period of the speaker in steps of 1 ms. Here, instead of considering two window sizes before and after the center window size, increasing or decreasing the number of window sizes was carried out. Increasing the number of window sizes makes the F1 score curve smoother, and decreasing the number of window size results in sudden fluctuations in the F1 score curve.

The evaluation of the detection performance was carried out using a leave-one-speaker-out strategy, where the speech data of a given speaker was held out for testing and the remaining speech data of all speakers was used for training. This procedure was repeated for each speaker. The proposed method is compared with two existing techniques:

1. **Ishi's method [8]:** In this method, short-term power, intraframe periodicity, and interpulse similarity measures are extracted to differentiate the creaky regions from modal and unvoiced regions.
2. **Kane-Drugman (KD) method [9]:** Creaky detection is performed by utilizing two acoustic features which are designed to characterize the presence of secondary peaks and prominent impulse-like excitation peaks from the LP residual signal.

F1 score, TPR, and FPR obtained for different speech databases are shown in Fig. 6.6. From the figure, it can be observed that the proposed method using epoch parameters performs better than the two existing methods, across all databases. Among all results, Ishi's method displays the lowest TPR and F1 score for MV speaker. The main reason for the lowest TPR and F1 score is that the MV speaker has modal regions at low frequency. Intraframe periodicity values extracted from Ishi's method dropped below a threshold value in low-voiced regions. Hence, the modal regions having low-frequency regions were wrongly identified as creaky regions (evident by high values of FPR). The proposed method produced a high F1 score and TPR of MV speaker, as the extracted variance of epoch parameters was independent of the pitch of the speech signal. KD method performed better than Ishi's method for all speech databases, but its performance is inferior, compared to the proposed method. One-way ANOVA is carried out to investigate whether the performance of the proposed creaky detection method is significantly better than the two existing methods. Here, F1 score is treated as the dependent variable and detection method as the independent variable. One-way ANOVA indicated that the creaky detection method had a significant effect on the F1 score [$F = 16.0$, $p < 0.001$], and pair-wise comparisons carried out using Tukey's honestly significant difference (HSD) test showed that the proposed method gave significantly higher F1 scores than both Ishi's method ($p < 0.001$) and KD method ($p < 0.01$).

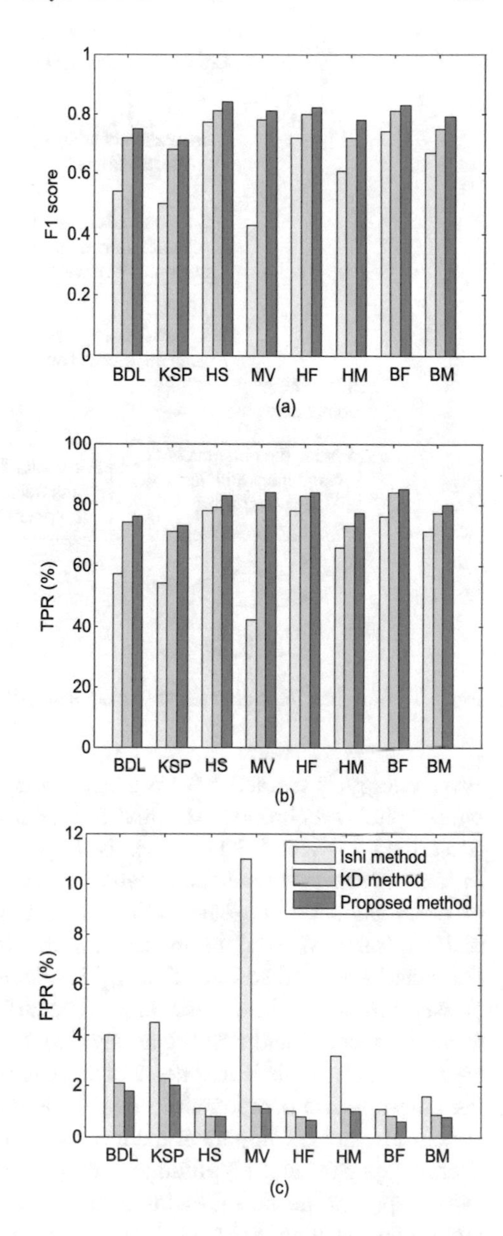

Fig. 6.6 Performance of three creaky detection methods on BDL, HS, MV, HM, HF, BM, and BF databases: (**a**) F1 score, (**b**) TPR, and (**c**) FPR

6.3 Hybrid Source Model for Generating Creaky Excitation

In this section, the proposed hybrid source model capable of generating the creaky excitation signal is described. Figure 6.7 shows the flow diagram of the proposed hybrid source model, which is an extension of the basic time-domain deterministic plus noise model (described in Chap. 5) for generating the creaky excitation.

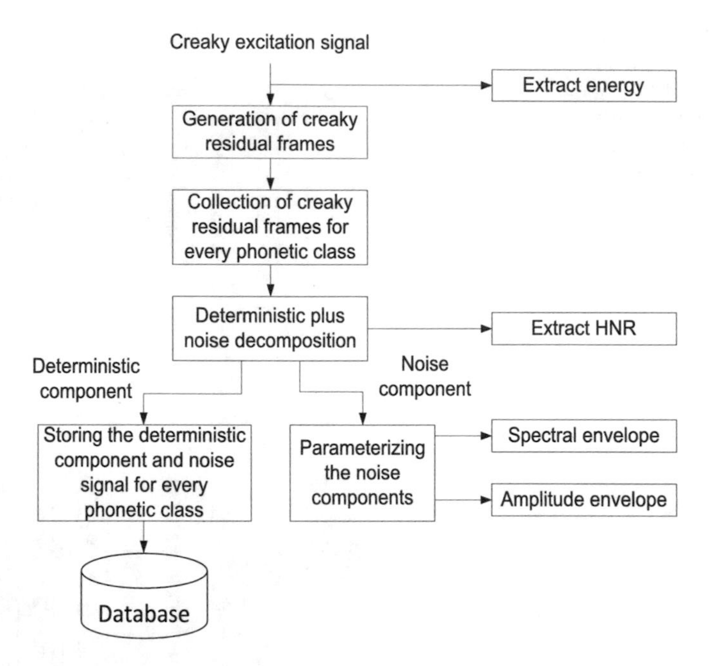

Fig. 6.7 Flowchart indicating the sequence of steps in the proposed creaky excitation modeling

First, energy is extracted from every frame of the creaky excitation signal. The creaky pitch-synchronous residual frames are extracted from the excitation signal (described in Sect. 6.3.1). Here, both glottal closure and secondary excitation instants of the creaky residual frames are synchronized. The creaky residual frames of every phonetic class are collected from the entire database. From the creaky residual frames of every phonetic class, the deterministic and noise components are computed (detailed in Sect. 6.3.2). The deterministic components of all phonetic classes are stored in the database. The noise components are parameterized in terms of spectral and amplitude envelopes. HNR is computed as the ratio of the energy of deterministic and noise components. For every phonetic class, along with the deterministic component, a single natural instance of the noise signal is also stored. The noise component having spectral and amplitude envelopes close to the average spectral and amplitude envelopes of all noise components is considered as the appropriate noise signal of a phonetic class. Energy, HNR, spectral, and amplitude envelopes are considered as creaky parameters and modeled under HMM framework. During synthesis, for generating the creaky excitation, the suitable creaky deterministic component of a phone is chosen from the database, and the noise component is obtained by imposing the target spectral and amplitude envelopes on the natural instance of noise signal (explained in Sect. 6.3.3). The issues involved in the modeling of the creaky excitation signal are explained in the following section.

6.3.1 Generation of Creaky Residual Frames

For proper source modeling, the pitch-synchronous residual frames should be extracted appropriately from the creaky regions. In the creaky voice, in addition to primary excitation at GCI, the secondary excitation peaks are also present [11]. Figure 6.8 shows the excitation signals of modal and creaky voices. In the figure, the GCI locations and secondary excitations are indicated by downward arrows (▼) and stars ($\star$), respectively. To obtain the appropriate creaky pitch-synchronous residual frames, accurate location of GCIs and secondary excitation instants are very essential. In this work, the zero frequency filtering method is used to extract optimal GCI locations in creaky regions. Secondary excitation peaks are detected by identifying the prominent peak between consecutive GCIs. The region around GCI (around $\pm$ 1 ms) is not considered to avoid detection in the close vicinity of GCI. Previous studies have reported that the secondary excitation peaks occur due to sharp discontinuities at the glottal opening instant, following a long glottal closed period [9, 11]. In this work, the time interval between the secondary excitation peak and its following GCI is termed as open period, and the time interval between the GCI and its subsequent secondary excitation peak is termed as closed period.

The procedure for extracting the pitch-synchronous residual frames from creaky regions differs from that of modal regions. In modal voice, the amplitude of the excitation signal is maximum at GCI. The pitch-synchronous residual frames are extracted in such way that the GCI is at the center of the frame and the length of the frame is two pitch periods. By ensuring GCI at the center of the frame results in the accurate estimation of deterministic and noise components. In the case of creaky excitation, in addition to GCI, the secondary excitation peaks of the creaky residual frames should also be synchronized. To synchronize the secondary excitation peaks, the open and closed periods of the creaky residual frames are analyzed. Figure 6.9 shows the distribution of open and closed periods of the creaky residual frames for

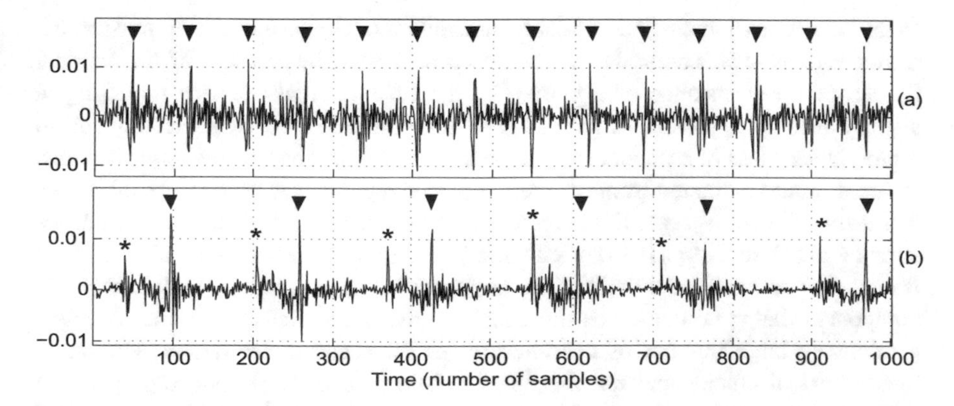

Fig. 6.8 Excitation signal of (**a**) modal voice and (**b**) creaky voice. GCI locations and secondary excitations are indicated by downward arrows (▼) and stars ($\star$), respectively

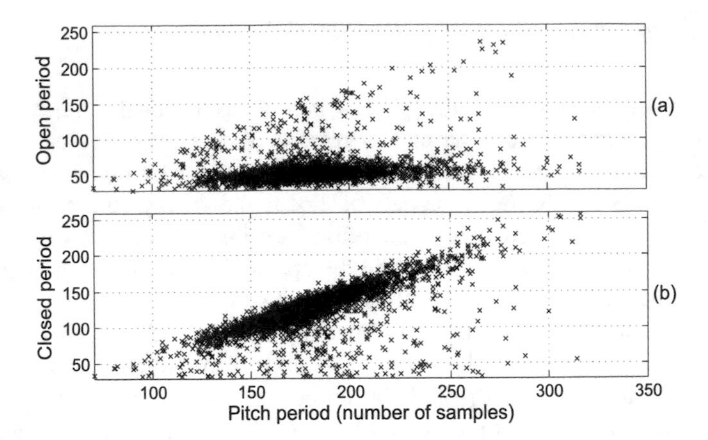

Fig. 6.9 Distribution of (**a**) open periods and (**b**) closed periods of creaky residual frames for the male speaker (BDL) obtained from CMU Arctic database

the male speaker (BDL) obtained from CMU Arctic database [5]. From the figure, it can be observed that the closed period varies almost linearly with the pitch period. The open period is narrowly distributed around a single value for different pitch periods. In this work, the open periods of the creaky residual frames are set to a constant value (3.75 ms for BDL speaker). The constant open period duration of all residual frames ensures the synchronization of secondary excitation peaks. The creaky residual frames are pitch-normalized by resampling only the closed period.

6.3.2 Deterministic Plus Noise Decomposition for Every Phonetic Class

In the basic time-domain deterministic plus noise model-based hybrid source model, the deterministic plus noise decomposition is performed by considering the residual frames of every instance of a phone. In the case of creaky source modeling, a slightly different approach is followed. The duration of the occurrence creaky voiced region is very less in a phone, and subsequently, the number of pitch-synchronous residual frames extracted from the creaky voiced regions is also very less (about 4–5 frames). On an average, the percentage of creaky regions in the speech corpus varies from 3 to 12% [10]. By considering a small number of creaky residual frames, the estimated deterministic component is not accurate. The deterministic component that is computed as the ensemble average of residual frames requires a relatively large number of residual frames for accurate estimation. Hence, in the context of creaky regions, the deterministic plus noise decomposition is not performed on every instance of a phone. Instead, the creaky residual frames of all phonetic classes are collected from the entire database, and the deterministic

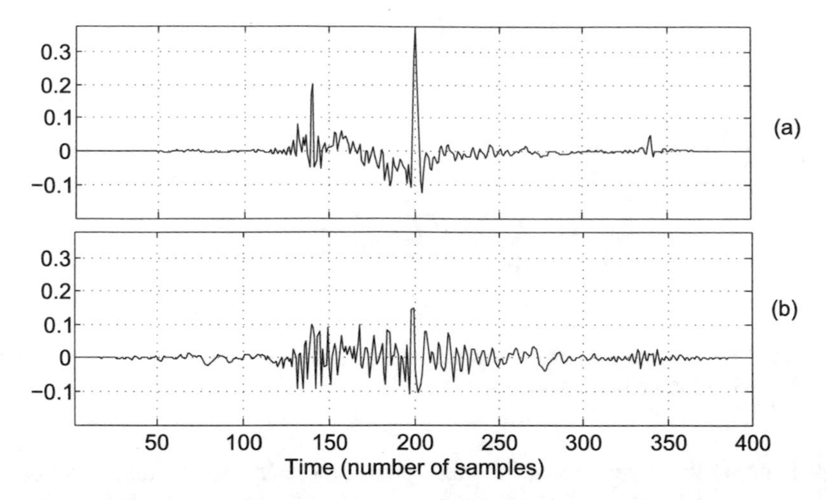

Fig. 6.10 (**a**) The deterministic component and (**b**) a single instance of noise component corresponding to the creaky residual frames of phone /aa/

plus noise decomposition is carried out separately for every phonetic class. In the BDL speaker of CMU Arctic database, 28 voiced phonetic classes are present. By considering the creaky residual frames of every phonetic class, the deterministic and noise components are computed. Figure 6.10 shows the deterministic component and a single instance of noise component corresponding to the creaky residual frames of phone /aa/.

The deterministic component computed for every phonetic class is stored in the database. The noise component is obtained by subtracting the deterministic component from the creaky residual frame of a particular phonetic class. For parameterizing the noise component, the same procedure is followed as described in parametric and hybrid source modeling methods. The noise component is parameterized in terms of its spectral and amplitude envelopes. The spectral envelope of the noise component is estimated by using 10th order LPC which are converted to LSF coefficients. The amplitude envelope of the noise component is represented by downsampling it into 15 samples. For modeling the parameters, the same approach is followed as described in parametric and hybrid source modeling methods.

6.3.3 Synthesis of Creaky Voice Using Proposed Hybrid Source Model

The block diagram showing different stages in the synthesis of creaky voice is presented in Fig. 6.11. In the figure, the parameters generated by the HMMs are shown in italics. During synthesis, depending on the generated creaky probability, each frame is classified as either creaky or non-creaky. For the creaky frame,

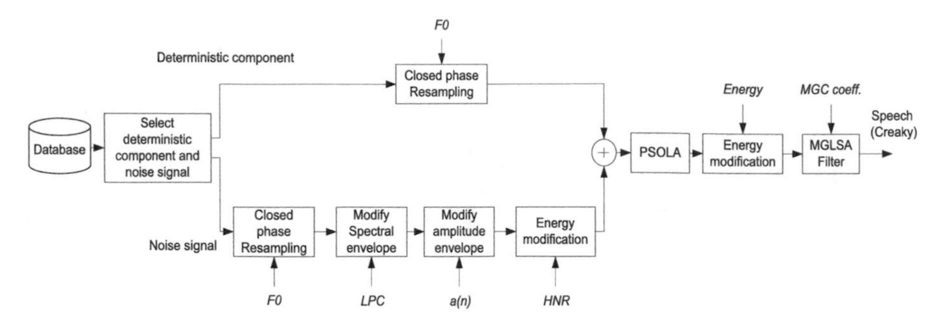

Fig. 6.11 Block diagram showing different stages in the synthesis of creaky voice. The parameters generated by HMMs are shown in italics

depending on the input phone, the corresponding creaky deterministic component and the natural instance of noise signal are selected from the database. The pitch period of the deterministic component and the natural instance of noise signal are transformed into target pitch period by resampling only the closed period. The target spectral and amplitude envelopes generated by the HMMs are imposed on the noise signal. For imposing the target spectral and amplitude envelopes on the noise signal, the same approach is followed as described in parametric and hybrid source modeling methods. After imposing the target spectral and amplitude envelopes, the energy of the noise signal is modified according to the generated HNR, and the energy of the deterministic component is kept constant throughout the phone. Both deterministic and noise components are pitch-synchronous overlap-add, and the gain of the combined signal is matched according to the energy measure generated by the HMM. The resulting excitation signal is given as input to the MGLSA filter, controlled by MGC coefficients to obtain the speech with creaky voice. For the modal voiced regions, the basic time-domain deterministic plus noise model-based hybrid source model is used for generating the excitation signal. For the unvoiced speech, the white noise whose energy is modified according to the generated energy measure is used as the excitation signal.

6.3.4 Evaluation

The proposed method is evaluated using two male speakers (BDL and KSP) from CMU Arctic speech database. The main reason for choosing BDL and KSP speakers for the evaluation is that they have the highest percentage of creaky regions among all other speakers in the CMU Arctic speech database. The amount of creaky regions present in the BDL and KSP speakers' databases are 7.6% and 2.7%, respectively (from Table 6.1). Each of the speakers' databases consists of 1100 phonetically balanced English utterances. The durations of BDL and KSP speakers' databases are 51 and 59 min, respectively. For every speech utterance, the corresponding

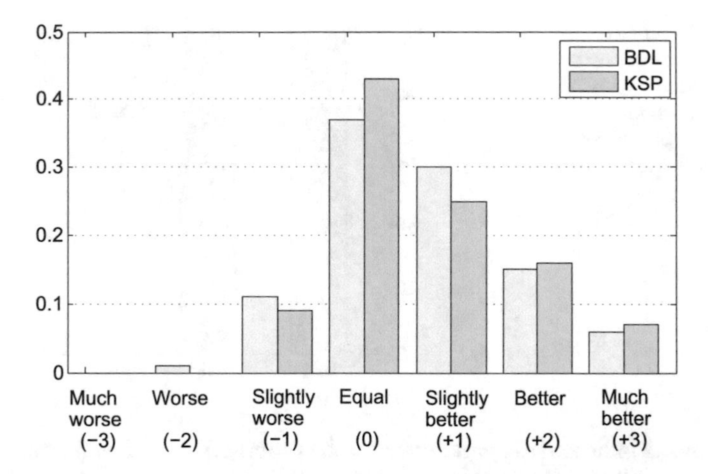

Fig. 6.12 Results of the CMOS test comparing the proposed method with the basic time-domain deterministic plus noise model-based source model

phonetic transcriptions are available in the CMU Arctic speech database. Based on the observation of the distribution of open periods for different pitch periods, the open periods of the creaky residual frames are set to 3.75 and 4.375 ms for BDL and KSP speakers, respectively. The proposed hybrid source model is compared with the basic time-domain deterministic plus noise model-based source model meant for generating only modal voice. The subjective evaluation is conducted with 20 research scholars having sufficient speech knowledge for proper assessment of the speech signals. The subjects were given a pilot test about the perception of creaky and non-creaky regions by playing the samples of synthesized speech files. Once they were comfortable with judging, they were allowed to take the tests. Initially, 100 sentences that were not part of the training data were synthesized using the proposed hybrid source model. Among 100 sentences, 20 synthesized speech files that have more than 3% of the creaky region are selected for the evaluation.

The subjective evaluation is carried out using two measures, namely, CMOS and preference tests. The results of the CMOS tests are shown in Fig. 6.12. The figure provides a comparison between the proposed hybrid source model and time-domain deterministic plus noise model-based source model according to the CMOS 7-point scale. In most of the cases, the subjects perceived that the proposed method is either *better* or *equivalent* to the time-domain deterministic plus noise model-based source model. The proposed method was never perceived as *much worse* than the time-domain deterministic plus noise model-based source model. The CMOS scores with 95% confidence interval obtained for BDL and KSP are 0.88 ± 0.021 and 0.93 ± 0.018, respectively.

In the preference tests, the subjects were asked to give the preference between a pair of synthesized speech utterances. The subjects had the option either to prefer one of the synthesized speech utterances or to prefer both as equal. The preference scores are provided in Fig. 6.13. From the figure, it can be observed that around 50%

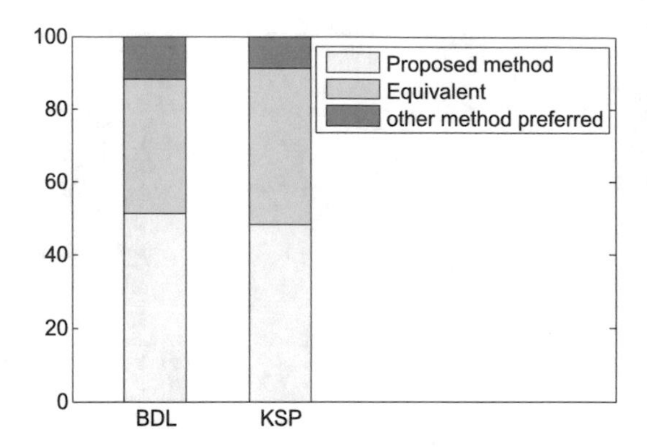

Fig. 6.13 Preference scores obtained by comparing the proposed method with the basic time-domain deterministic plus noise model-based source model

of the subjects indicated the preference for the proposed method and around 10% of the subjects preferred the basic time-domain deterministic plus noise model-based source model compared to the proposed method. Finally, around 40% of the subjects identified both methods as equal. These results confirm that the incorporation of creaky excitation has improved the quality of HTS.

Synthesized speech samples of the proposed hybrid source model (for generating creaky and modal voice) and the basic time-domain deterministic plus noise model-based hybrid source model (for generating only modal voice) are made available at http://www.sit.iitkgp.ernet.in/~ksrao/narendra-phd-demo/narendra_hsm.html.

6.3.5 Discussion

From the subjective evaluation results, it is observed that the utilization of creaky deterministic component particular to every phonetic class resulted in the improvement of the quality of HTS. For two male speakers used in the evaluation, two separate HTS are developed. As single speaker data is used for developing the synthesis system, speaker normalization is not performed during training and synthesis stages of HTS, and hence, mean F_0 information is not utilized in HTS. The proposed creaky voice detection method accurately identified the creaky regions in the speech utterances which prompted generation of creaky voice at appropriate places in the synthesized speech. Voicing detection and F_0 estimation are performed using a recently proposed method based on the strength of excitation. Using this method, most of the creaky regions are detected as voiced, and accurate F_0 values are extracted in creaky regions (without F_0 halving and doubling errors). From informal listening tests of different speech utterances containing creaky voice, the following observations can be drawn. The creaky regions are mostly occurring at the end of the phrase or utterance where energy contour is tapering down. If the creaky frames of a phone are less than 10% the length of a phone, then the creaky segments are not perceived in most of the cases. As the length of creaky segments is small compared to entire speech utterance, it is difficult to perceive the effectiveness of the creaky segments at the speech utterance level. But, when the speech utterance is observed in small chunks, then the perceptual significance of the creaky regions can be noticed.

As both modal and creaky regions are generated within a phone, there may exist slight distortion and F0 fluctuations at the joining portions of modal and creaky regions in the synthesized speech. At certain places where the creaky regions are wrongly synthesized in place of modal voiced regions, slight unnaturalness can be perceived in the synthesized speech. In our database, most of the creaky regions contain secondary excitation. The hybrid source model is proposed by implicitly assuming that all the creaky residual frames contain secondary excitation. Hence, the proposed method cannot be directly applied on the speakers who do not produce secondary excitation in the creaky regions. The proposed creaky voice detection and hybrid source modeling method are performed by analyzing the segments of the creaky voice of neutral speaking style speaker. The proposed method should be suitably modified to be applied for other speaking styles such as storytelling, interactive conversation, and expressive speech.

6.4 Summary

In this chapter, two issues involved in the generation of creaky voice in the HMM-based speech synthesis system were addressed. First, the creaky voice detection method was proposed based on the variation of epoch parameters for different voicing regions. The neural network classifier was trained using the variance of epoch parameters obtained from the ZFF method. The hybrid source model was proposed to generate the creaky voice at appropriate places in the synthesized speech. In the proposed hybrid source model, the deterministic plus noise decomposition was performed on the creaky residual frames of every phonetic class. During synthesis, the suitable creaky deterministic component was obtained from the database, and the noise component was generated from the parameters generated by the HMMs. The creaky deterministic and noise components were pitch-synchronous overlap-added to generate the creaky excitation signal. Both CMOS and preference tests indicated that the incorporation of creaky excitation has improved the quality of HTS compared to the basic time-domain deterministic plus noise model-based hybrid source model.

References

1. K.S.R. Murty, B. Yegnanarayana, Epoch extraction from speech signals. IEEE Trans. Audio Speech Lang. Process. **16**(8), 1602–1613 (2008)
2. K.S.R. Murty, B. Yegnanarayana, M.A. Joseph, Characterization of glottal activity from speech signals. IEEE Signal Process. Lett. **16**(6), 469–472 (2009)
3. P. Alku, T. Bakstrom, E. Vikman, Normalized amplitude quotient for parameterization of the glottal flow. J. Acoust. Soc. Am. **112**(2), 701–710 (2002)

4. C.M. Bishop, *Pattern Recognition and Machine Learning* (Springer, Berlin, 2006)
5. CMU ARCTIC speech synthesis databases [Online]. http://festvox.org/cmu_arctic/
6. H. Sil, E. Helander, K. Koppinen, M. Gabbouj, Building a Finnish unit selection TTS system, in *Proceedings of International Speech Communication Association Speech Synthesis Workshop 6 (ISCA SW6)* (2007), pp. 310–315
7. M. Vainio, Artificial neural network based prosody models for Finnish text-to-speech synthesis. Ph.D. dissertation, University of Helsinki, Finland, 2001
8. C. Ishi, K. Sakakibara, H. Ishiguro, N. Hagita, A method for automatic detection of vocal fry. IEEE Trans. Audio Speech Lang. Process. **16**(1), 47–56 (2008)
9. J. Kane, T. Drugman, C. Gobl, Improved automatic detection of creak. Comput. Speech Lang. **27**(4), 1028–1047 (2013)
10. T. Drugman, T. Raitio, Excitation modeling for HMM-based speech synthesis: breaking down the impact of periodic and aperiodic components, in *Proceedings of International Conference on Audio, Speech and Signal Processing (ICASSP)* (2014), pp. 260–264
11. M. Blomgren, Y. Chen, M. Ng, H. Gilbert, Acoustic, aerodynamic, physiologic, and perceptual properties of modal and vocal fry registers. J. Acoust. Soc. Am. **103**(5), 2649–2658 (1998)

Chapter 7
Summary and Conclusions

7.1 Summary of the Book

In this work, we addressed different issues related to the modeling of the excitation signal. One of the important excitation features which greatly influence the quality of synthesis is pitch or F0. The existing F0 estimation methods' performance degrade sharply in creaky or low-voiced regions. Hence, we have proposed a method for accurate voicing detection and F0 estimation in modal and low-voiced regions [1]. Impulse-like excitation present in the voiced speech is utilized for extracting the fundamental frequency. Zero frequency filter is used to derive the locations of impulse excitation. The size of the window used for computing the local mean in the ZFF method is explored for accurate voicing detection and F0 estimation. The strength of excitation is efficiently computed from the ZFF method with different window sizes. By adaptively choosing appropriate window size, the strength of excitation for the voiced speech is significantly higher compared to the unvoiced speech. With the suitable threshold on the strength of excitation, accurate voicing detection is performed. With the appropriate window size, the speech signal is passed through the zero frequency filter to derive smooth and accurate F0 contour. The performance of the proposed method is compared with other existing voicing detection and F0 estimation methods. The results indicate that the proposed method resulted in significant improvement of the voicing detection. It is also observed that the gross pitch error has reduced, which indicates that the F0 halving or doubling errors have reduced significantly. The proposed voicing detection and F0 estimation method is implemented in HMM-based speech synthesis system. The evaluation of voicing detection in the synthesized speech showed a remarkable reduction in voiced/unvoiced decision errors for the proposed method compared with RAPT and STRAIGHT F_0 estimation methods. The subjective evaluation performed using CMOS and preference tests indicate that the speech synthesized using the proposed

K. S. Rao, N. P. Narendra, *Source Modeling Techniques for Quality Enhancement in Statistical Parametric Speech Synthesis*, SpringerBriefs in Speech Technology,
https://doi.org/10.1007/978-3-030-02759-9_7

method is better than the speech synthesized using the RAPT and STRAIGHT F_0 estimation methods.

For the accurate generation of excitation signal during synthesis, the excitation signal is modeled using parametric and hybrid approaches. In the parametric approach, two methods are proposed for modeling the pitch-synchronous residual frames of the excitation signal based on the principal component analysis. In the first method, based on the evolution of cumulative relative dispersion for different eigenvectors, the pitch-synchronous residual frames are parameterized using 30 PCA coefficients. In the second method, an analysis of characteristics of the residual frames around GCI is performed using PCA [2]. Based on the analysis, the segment of the residual signal around GCI is considered as the deterministic component, and the remaining part of the residual signal is considered as the noise component. The deterministic component is compactly represented using PCA coefficients, and the noise component is parameterized in terms of spectral and amplitude envelopes. During synthesis, the deterministic and noise components are reconstructed from the parameters generated by HMMs. The subjective evaluation results showed a significant improvement of the quality for both female and male speakers' speech synthesized by the proposed method, compared to three existing excitation modeling methods.

In the hybrid approach, two methods are proposed for modeling the excitation signal. In the first method, the excitation signal of a phone is modeled using a single optimal residual frame. The optimal residual frames extracted from the excitation signals of all phones are efficiently arranged in the form of a decision tree. During synthesis, using the suitable optimal residual frame selected from the decision tree, the excitation signal of a phone is reconstructed. Using a single optimal residual frame, cycle-to-cycle variations present in the residual frames of the excitation signal cannot be captured accurately. Hence, to accurately represent each of the residual frames and incorporate cycle-to-cycle variations in the residual frames of the excitation signal, the time-domain deterministic plus noise model-based hybrid source model is proposed [3]. In the proposed hybrid source model, each of the residual frames of the excitation signal is viewed as a combination of deterministic and noise components. The deterministic component is constant for all residual cycles of a phone, and the noise component varies for every residual cycle of a phone. The deterministic components estimated from all phones are systematically arranged in the form of a decision tree. The noise components are parameterized in terms of spectral and amplitude envelopes. During synthesis, for the given input, the appropriate deterministic component is selected from the decision tree, and the noise components are obtained from the parameters generated by HMMs. The excitation signal of a phone is generated by combining the deterministic and noise components. The speech synthesized using the proposed source model is compared with three existing source modeling methods. Both CMOS and preference tests have indicated that the proposed method is better than three existing source modeling methods.

In order to produce good quality speech, HMM-based speech synthesis system should be capable of producing the creaky voice, in addition to modal voice. Most of the source modeling approaches are optimized to generate the speech particular to modal phonation. In this work, an HMM-based speech synthesis system capable of generating the creaky voice in addition to modal voice is developed. During the training phase, the creaky regions are identified from the speech utterance. The detection of creaky voice is performed based on the analysis of the variation of epoch parameters for different voicing regions. Using the variance of epoch parameters, a neural network classifier is trained to detect the creaky regions. The performance evaluation results indicate that the proposed method performs significantly better than the two existing creaky voice detection methods on different speech databases. During synthesis, to generate the creaky voice at appropriate places in the synthesized speech, the hybrid source model is proposed [4]. The hybrid source model is an extension of previously proposed time-domain deterministic plus noise model-based source model. In the proposed hybrid source model, the deterministic and noise components are computed from the creaky residual frames of every phonetic class. The deterministic components of every phonetic class are stored in the database, and the noise components are parameterized in terms of spectral and amplitude envelopes. During synthesis, every frame of a phone is classified as creaky/non-creaky based on the creaky detection probability. For every creaky frame of a phone, the corresponding creaky deterministic component is selected from the database, and the noise components are obtained from the parameters generated from the HMMs. The creaky excitation signal is constructed by combining the deterministic and noise components.

7.2 Contributions of the Book

- A robust voicing detection and F_0 estimation method is proposed based on the ZFF method.
- A parametric source modeling method is proposed based on the principal component analysis of pitch-synchronous residual frames of the excitation signal.
- Based on the analysis of characteristics of the residual frames around GCI, a parametric source modeling method is proposed which models the residual frames as a combination of deterministic and noise components.
- A hybrid source modeling method is proposed by utilizing the optimal residual frames extracted from the excitation signal of a phone.
- A hybrid source modeling method is proposed based on the time-domain deterministic plus noise model.
- Automatic detection of creaky voice method is proposed based on the variance of the epoch parameters. The epoch parameters which characterize the source of excitation are analyzed in modal and creaky regions.
- The hybrid source model is proposed for generating the creaky voice at appropriate places in the synthesized speech.

7.3 Directions for Future Work

- The accuracy of proposed voicing detection and F0 estimation method can be further improved by combining with other methods which exploit the periodicity of speech signals and the correlation properties of glottal cycles.
- Efficiency of the proposed voicing detection and F0 estimation method can be systematically analyzed for different voicing qualities such as modal, tensed, creaky, rough, breathy, and laughing.
- In the parametric source modeling approach, PCA analysis is performed by considering the residual frames of all phones. Instead, PCA analysis can be performed on the residual frames of every phonetic class, and improvement in the quality of the synthesized speech can be analyzed.
- Time-domain deterministic plus noise model-based hybrid source model is used for generating the excitation signal of the voiced phones. The proposed source model can be extended to produce the excitation signal of the unvoiced phones.
- In the proposed hybrid source model, the deterministic plus noise modeling is performed at the phone level. Additionally, it can be carried out at syllable or word level, and the evaluation of their performances at different levels can be conducted.
- The proposed source modeling approaches (both parametric and hybrid) can be implemented in expressive speech synthesis system, and the evaluation of improvement in the quality of the synthesized expressive speech can be performed.
- The proposed source modeling methods perform deterministic plus noise decomposition on the time-domain representation of the excitation signal. For the better understanding of different source modeling methods, the relation between the time and frequency-domain decomposition of the excitation signal can be analyzed.
- In this work, primary analysis of the dependency of the excitation signal on different phonetic classes is performed. Extensive analysis of the dependency of the excitation signal on the phones and its context can be performed for improved modeling of the source signal.
- The phase information of the residual frame can be explored for enhancing the performance of the proposed source modeling method.

References

1. N.P. Narendra, K. Sreenivasa Rao, Robust voicing detection and F0 estimation for HMM-based speech synthesis. Circuits Syst. Signal Process. **34**(8), 2597–2619 (2015)
2. N.P. Narendra, K. Sreenivasa Rao, Parameterization of excitation signal for improving the quality of HMM-based speech synthesis system. Circuits Syst. Signal Process. **36**(9), 3650–3673 (2017)

3. N.P. Narendra, K. Sreenivasa Rao, Time-domain deterministic plus noise model based hybrid source modeling for statistical parametric speech synthesis. Speech Commun. **77**, 65–83 (2016)
4. N.P. Narendra, K. Sreenivasa Rao, Generation of creaky voice for improving the quality of HMM-based speech synthesis. Comput. Speech Lang. **42**, 38–58 (2017)

Index

K. S. Rao, N. P. Narendra, *Source Modeling Techniques for Quality Enhancement in Statistical Parametric Speech Synthesis*, SpringerBriefs in Speech Technology,
https://doi.org/10.1007/978-3-030-02759-9

Printed in the United States
By Bookmasters